FOCUS ON BIOTERRORISM

FOCUS ON BIOTERRORISM

GEORGE T. FOSTER
EDITOR

Nova Science Publishers, Inc.
New York

For permission to use material from this book please contact us:
Telephone 631-231-7269; Fax 631-231-8175
Web Site: http://www.novapublishers.com

Library of Congress Cataloging-in-Publication Data

Focus on bioterrorism / George T. Foster (editor).
p. cm.
Includes index.
ISBN 1-60021-185-2
1. Biological warfare. 2. Bioterrorism. 3. War on Terrorism, 2001- I. Foster, George T.
UG447.8.F63 2006
363.325'3--dc22 2006013134

Published by Nova Science Publishers, Inc. ✦New York

CONTENTS

PREFACE

The anthrax murders in the U.S. and discoveries of biological weapon plans in Africa served as a wake-up call (too late as usual) that biological weapons might actually be a weapon of mass destruction. Their small size but highly deadly nature adds to the fear factor. The reality of this potential danger to the U.S. population found both the federal government as well as state government clueless. This new book presents analyses of the actions planned or already carried out in the battle against the threat of bioterrorism, and offers frank analyses of our current state of preparedness or lack thereof.

Attenuated vaccinia virus (VV) strains LC16, LC16mO (mO) and LC16m8 (m8) were developed from LO strains as temperature-sensitive variants by using cultured rabbit kidney cells in Japa, as duscussed in chapter 1. Among them m8 showed less pathogenicity compared with other strains, and was vaccinated into more than 50,000 children in Japan during 1974-1975 without inducing serious side effects. The m8 strain is the first licensed cell culture smallpox vaccine in the world that forms small plaques due to a frame-shift in the *B5R* gene coding an envelope protein. Recently, the Japanese Government has used the m8 strain as a stockpile virus prepared against bioterrorism. Here, we review m8 and mO strains from the following points: 1) history, 2) recent progress in molecular genetics, including *B5R* gene responsible for attenuation, 3) immunological and pathological properties, and 4) application as a vector of recombinant vaccines.

On June 20, 2002, an advisory panel to the U.S. Centers for Disease Control and Prevention (CDC) voted against mass smallpox vaccination and instead recommended that 10-20 thousand first responders receive the vaccine as reported in chapter 3. The recommendations are currently under review by CDC and the Department of Health and Human Services (HHS). Congress addresses control of the vaccine stockpile in the Homeland Security legislation (H.R. 5005, S. 2452).

The deliberate mailing of *Bacillus anthracis* spores through the U.S. Postal Service (USPS) has caused five deaths, twenty-two cases of anthrax, and massive disruptions to Congress and the USPS, as discussed in chapter 3.

Both the public and private sector are examining an array of methods to limit the risk of future attacks. The array of potential solutions include improving mail handling procedures, changing the USPS anonymous mailing policy, installing bio/chem agent detectors, and sterilizing the mail.

For the USPS these decisions are complicated by its precarious financial state. Some proposed solutions may require an increase in postage rates and/or decreased levels of service. Each of these may further depress postal revenues and threaten the continued existence of the USPS as an independent, self-supporting entity.

Policymakers may wish to consider if the USPS must ensure the safety of mail recipients. At this point it is not clear if this is practical or even possible with existing technology. It may be that it is practical to protect only the mail addressed to the most likely targets of future attacks.

Some of the measures that the USPS has taken or is planning to take to protect postal workers and mail recipients are common sense alterations to the mail processing procedures. These include measures to reduce cross contamination such as using vacuuming instead of pressurized air to dislodge dust and adding filters to air handling systems. The USPS is studying the feasibility of including biological weapon detectors during the sorting process. Additionally, the USPS has made gloves, masks, and educational materials available to all postal workers.

More controversial and potentially more costly are plans to sterilize the mail. Currently, all mail destined for federal offices in the Washington DC metropolitan area is shipped to a sterilization facility for irradiation treatment before delivery. The USPS is studying whether this solution can be scaled up to sterilize all mail from anonymous senders. To implement irradiation procedures nationwide could cost up to $2.25 billion with up to another billion dollars each year in operating costs. This procedure will cause delays and may damage the contents of some mail.

Also controversial and costly are plans to increase product tracking systems that will include video images of all retail transactions. The planned system will enable mail dropped off at post offices to be tracked through to delivery and linked to facial images of senders. Mail dropped off at anonymous collection sites will be traceable to the drop off point. Stamps and envelopes could also be encoded with purchase location information.

Policymakers face a difficult balance to strike between concerns for safety, cost, and practicality while deciding how to alter the practices of the USPS.

Chapter 4 reviews that discovery of anthrax spores in mail arriving at media and congressional offices in October 2001 changed the relationship of the U.S. Postal Service (USPS) to its customers and to Congress. In the immediate aftermath of the discovery, delivery of mail to 137 million addresses each day became a source of fear to some Americans and of uncertainty to many more. No institution was affected more directly than Congress itself, where dozens of staff members tested positive for anthrax exposure and thousands were barred from their offices. The use of mail as an instrument of bioterrorism now has Congress questioning whether the largely independent Postal Service has the capability and resources to be in the front lines of homeland defense. Already in deep financial trouble, USPS is looking to Congress for substantial financial assistance, and a fundamental change in its role in government is back on the agenda.

On September 11, 2001 the lives of many American citizens changed in an unexpected way, as discussed in chapter 5. The acts of terrorism that took place on that date caused us to become more conscious of our surroundings. Terrorism was brought into our lives in a different way, we were not watching it unfold on the news in another country, but were watching it unfold in the United States. Foreign terrorists had become a part of our society, they had moved into our neighborhoods and had become the ones that succeeded in destroying our sense of security. The acts of terrorism that took place on September 11 caused us to face terrorism once again at home. With the concern of terrorism still on the minds of many the concern soon changed to bioterrorism when letters containing anthrax began to appear in the mail, closing two major postal facilities on the east coast. On October 4, 2001 a Florida man was diagnosed with anthrax, he died on October 5, 2001 making him the first person in the United States to die as a result of a bioterrorist act.

The use of biological weapons has been attempted throughout history as a means to terrorize the enemy during wartime or as a public threat during times of peace. The use of biological pathogens has been well documented with bacteria, viruses, fungi or toxins, being used on poison arrows or to bomb an area with deadly bacteria and can be traced to as early as 190 BC by Hannibal who hurled venomous snakes onto the enemy ships of Pergamus at Eurymedon to panic and injure enemy sailors. The concern regarding the use of biological weapons remains throughout the United States and in foreign countries.

This chapter will focus on the historical aspects of bioterrorism tracing their use over the centuries to the current day. It will include a discussion of events that took place during World War I and World War II, related to the use of biological weapons and the decision made by President Nixon to discontinue offensive biological research and destroy the United States' arsenal.

The author of chapter 6 reviews that the use of sanctions specifically to stem weapons proliferation is a relatively new development in U.S. foreign policy. While earlier legislation required the cutoff of foreign aid to countries engaged in specified nuclear proliferation activities and mentioned other sanctions as a possible mechanism for bringing countries into compliance with goals of treaties or international agreements, it was not until 1990 that Congress enacted explicit guidelines for trade sanctions related to missile proliferation. In that year a requirement for the President to impose sanctions against U.S. persons or foreign persons engaging in trade of items or technology listed in the Missile Technology Control Regime Annex (MTCR Annex) was added that year to the Arms Export Control Act and to the Export Administration Act of 1979. Subsequently, Congress legislated economic sanctions against countries that contribute to the proliferation of chemical, biological, and nuclear weapons in a broad array of laws.

This chapter offers a listing and brief description of legal provisions that require or authorize the imposition of some form of economic sanction against countries, companies, or persons who violate U.S. nonproliferation norms. For each provision, information is included on what triggers the imposition of sanctions, their duration, what authority the President has to delay or abstain from imposing sanctions, and what authority the President has to waive the imposition of sanctions.

The former Soviet and subsequently Russian biological weapons program possessed capabilities far in excess of any such program known to have existed elsewhere, as discussed in chapter 7. These capabilities included genetically-altered, antibiotic-resistant pathogens and sophisticated delivery systems. Approximately fifty biological research and production centers (BRPCs) throughout the former Soviet Union devoted either all or part of their work to the program. In the post-Soviet era, former Soviet states drastically reduced and in some cases eliminated funding for these BRPCs. Thousands of BW scientists became unemployed or underemployed, and the facilities, weapons technology, and thousands of strains of pathogens at these BRPCs became vulnerable to theft, sale or misuse.

In the mid-1990s, the United States began engaging BRPCs throughout the former Soviet Union in four kinds of cooperative projects aimed at preventing proliferation of BW capabilities. Collaborative research projects involve former BW scientists in projects with

American scientists and seek to deter former BW scientists from selling their expertise to terrorist groups or proliferating states. Several U.S. government agencies are involved in collaborative research projects, most of which are funded through the international science centers. Biosafety enhancement projects are intended to make BRPCs safe places for collaborating scientists to work. In combination, collaborative research and biosafety enhancement projects give U.S. officials routine access to laboratories and facilities that were once used for BW research and production. BioSecurity projects consolidate and restrict access to pathogens. Dismantlement projects target excess infrastructure and BW equipment at BRPC sites for permanent dismantlement. Biosafety, BioSecurity, and Dismantlement projects are funded through and carried out by DOD's Cooperative Threat Reduction Program.

U.S. participants in these projects identify several lessons learned in the past few years. First, it has become clear that the infrastructure of the Soviet/Russian BW complex was more extensive than most analysts realized when the United States initiated its efforts to prevent proliferation of BW capabilities from former Soviet states. Cooperative projects at some BRPCs have helped open doors to other BRPCs, and since 1995, more than forty BRPCs have been involved in cooperative projects with the United States. Second, U.S. participants report that biosafety, biosecurity, and dismantlement projects require complex negotiations, complex engineering work, considerable project management support, and innovative solutions for problems specific to each BRPC. Consequently, they have learned that the United States may need to offer a long-term commitment if it wants to complete the effort. At the same time, the U.S. agencies with BW nonproliferation programs recognize the need to maximize the nonproliferation benefits of U.S. assistance in an environment with limited resources. Finally, U.S. participants have discovered that interpersonal and institutional relationships resulting from these cooperative efforts may play a powerful role in preventing proliferation of BW capabilities from former Soviet states.

In: Focus on Bioterrorism
Editor: George T. Foster, pp. 1-15

ISBN 1-60021-185-2

Chapter 1

PROPERTIES OF ATTENUATED VACCINIA VIRUS STRAIN LC16M8 STOCKPILED AGAINST BIOTERRORISM

Masanobu Sugimoto[1,*], Kazue Ohishi[2] and So Hashizume[3]
[1]Gene Care Research Institute, Co., Ltd., Kamakura, Kanagawa, Japan
[2]Japan Agency for Marine-Earth Science and Technology, Yokosuka, Kanagawa, Japan
[3]Japan Poliomyelitis Research Institute, Higashimurayama, Tokyo, Japan

ABSTRACT

Attenuated vaccinia virus (VV) strains LC16, LC16mO (mO) and LC16m8 (m8) were developed from LO strains as temperature-sensitive variants by using cultured rabbit kidney cells in Japan. Among them m8 showed less pathogenicity compared with other strains, and was vaccinated into more than 50,000 children in Japan during 1974-1975 without inducing serious side effects. The m8 strain is the first licensed cell culture smallpox vaccine in the world that forms small plaques due to a frame-shift in the *B5R* gene coding an envelope protein. Recently, the Japanese Government has used the m8 strain as a stockpile virus prepared against bioterrorism. Here, we review m8 and mO strains from the following points: 1) history, 2) recent progress in molecular genetics, including *B5R* gene responsible for attenuation, 3) immunological and pathological properties, and 4) application as a vector of recombinant vaccines.

Keywords: bioterrorism, attenuated vaccinia virus, LC16m8, smallpox, variola, stockpile vaccine

* Correspondence concerning this article should be addressed to Masanobu Sugimoto, Ph.D., GeneCare Research Institute, 200 Kajiwara, Kamakura, Kanagawa 247-0063, Japan. Tel: +81-467-46-9590, Fax: +81-467-48-6595 e-mail: sugimoto@genecare.co.jp.

INTRODUCTION

Smallpox used to be one of the most feared infectious diseases of humans. Thus, the success of smallpox eradication by the World Health Organization (WHO) program in the 1970s is a landmark in the history of biomedical science. WHO declared global eradication of smallpox in 1980 [1]. In the program, vaccinia virus (VV) vaccination was the most effective preventive measure against smallpox caused by variola virus, an orthopoxvirus [2].

The VV vaccine can produce serious complications, such as generalized eczema, progressive vaccinia and post-vaccinal encephalitis. The vaccine strains, the Lister and New York City Board of Health (NYBH), were shown to be less pathogenic of effective vaccines. However, according to data before eradication in 1968, the occurrence of serious complications, such as encephalitis, was about 20 cases for every 100 million people vaccinated [3,4]. For these reasons, together with the eradication of smallpox, WHO recommended stopping the vaccination in 1980. Consequently, all nations stopped their routine smallpox vaccination programs thereafter.

Since the incidents of September 2001 in the U.S.A., concern about bioterrorism has risen. Among many pathogens that could potentially be used in bioterrorism, variola virus is one of the most recognized and the most feared. Thus, after a lag time of more than 20 years, serious attempts have been made to restart the development of lower-virulence vaccine strains, including a vaccinia ACAM1000 clone established from Dryvax (NYBH strain) [5], modified vaccinia Ankara (MVA) and NYVAC (NYBH strain) replication-incompetent viruses [6,7]. However, the vaccine strains deriving NYBH, including ACAM and Dryvax vaccines, may have a side effect of inducing myocarditis [8,9], and replication-incompetent viruses may have the potential problem of relatively poor immunogenicity.

In the early 1970s, Hashizume *et al.* [10,11], then at Chiba Serum Institute, Japan, developed one of the safest replication-competent vaccines, vaccinia LC16m8 (m8) strain, from the Lister (Elstree) original strain (LO) that was used worldwide in the WHO program. The m8 strain is now stockpiled as a smallpox vaccine against bioterrorism in Japan: the Japanese government stockpiled m8 vaccine for 2,500,000 people at the end of March, 2002 and planned to increase the amount of the stockpile [12].

In this review, we describe m8 and its related variant strain, LC16mO (mO), from the following points: 1) history, 2) recent progress in molecular genetics, including the *B5R* gene responsible for attenuation, 3) immunological and pathological properties, and 4) application as a vector of recombinant vaccines.

History of m8

In the WHO eradication program of the 1970s, all vaccines used, including the LO strain, were produced in the skin of calves and sheep: VV was inoculated in the disinfected skin of animals, and it was recovered from the lymph. This first-generation vaccine has disadvantages: 1) a risk of contamination with microorganisms, including bacteria, and 2) difficulty in controlling adventitious viruses. Also, the first-generation vaccines induce encephalitis as a side effect.

To solve these problems, attenuated VV strains LC16, mO and m8 were developed by Hashizume *et al.* [10,11] from LO strains by using cultured rabbit kidney cells. Table 1 shows a schematic process to establish LC16, mO and m8.

Table 1. Establishment of LC16, LC16mO and LC16m8 strains.

Virus strains	Selection process	Markers for selection
Lister (Elstree) virus (LO)		
↓	36 passages in RK cells at 30 °C	Ts: not replicating over 40.8 °C
LC16		
↓	6 passages in RK cells at 30 °C	Ts, and pock size on CAM
LC16mO (mO)		
↓	3 passages in RK cells at 30 °C	Pock size on CAM
LC16m8 (m8)		

Ts, temperature sensitivity: RK, rabbit kidney: CAM, chorioallantoic membrane.

LC16 Strain

The LC16 strain was first selected by passaging the LO strain 36 times in primary rabbit kidney cells at a low temperature (30 °C), and were selected by temperature sensitivity: one clone, designated LC16, showed the lowest titer at 40.8 °C among the cloned viruses. Compared with LO strain, the LC16 strain had slightly increased growth in rabbit skin, but had a much reduced growth in the central nervous system of rabbit and monkey as compared with that of highly attenuated strain DIs [11].

The most conspicuous feature of the LC16 strain is a much lower neurovirulence in monkeys and rabbits than the parental LO strain or other vaccine strains used in the eradication program [11,13]. For instance, neurovirulence of LC16 in monkeys is comparable to that of DIs that lacks proliferating ability in humans. Comparative studies of the virulence of several VV strains, including the LC16 strain, by intrathalamic inoculation into cynomolgus monkeys showed the following results (Figure 1): the CV1 virus was most virulent; the NYBH, Ikeda, EM63, and Lister viruses were slightly less virulent; and DIs and LC16 viruses were least virulent. Characteristic findings were widespread inflammatory lesions in the meninges and choroid plexus that were closely associated with the replication of VV, and parenchymal lesions that might be identified as encephalopathy in deceased monkeys.

mO and m8 Strains

A mutant that had a lower ability to grow in skin was selected by using a pock size marker on the chorioallantoic membrane (CAM) of embryonated eggs. After six passages of the LC16 virus in primary rabbit kidney cells at 30 °C, a single plaque that produced medium-size pocks (2-3 mm) was isolated and was named LC16mO (mO). After three passages of mO virus, a small pock-forming clone (0.5 – 1.0 mm), m8, was isolated [11] (Figure 2). The strain m8 showed lower replication activity in Vero cells compared with mO and LC16 strains:

virus titers of m8 were two logs lower in Vero cells than mO and LC16. The titers in chick embryonic fibroblasts differed little between m8, mO and LC16 strains. Thus, the low ability to grow in Vero cells is a notable property of m8.

Table 2 summarizes the properties of LO and its attenuated variants, LC16mO, LC16m8 and LOTC. For LOTC, see the section "THE *B5R* GENE".

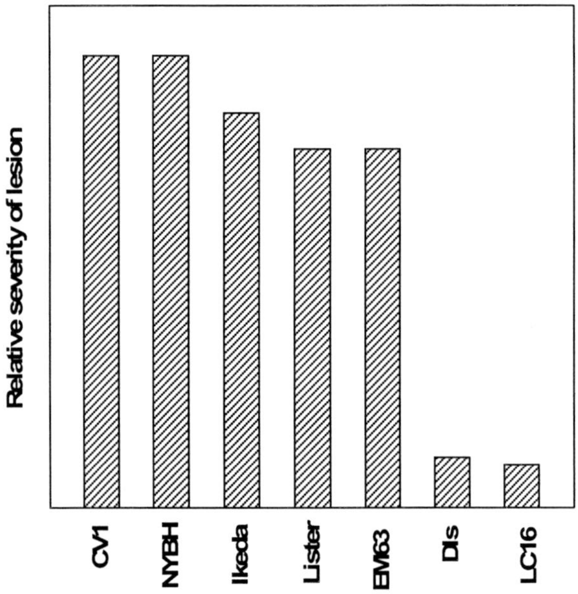

Figure 1. Severity of inflammatory lesions in the meninges and choroid plexus in monkeys. The vertical axis indicates the relative severity of the lesion, and the VV strains tested are shown on the horizontal axis. Monkeys were inoculated with 2 x 10^7 plaque-forming units of the VV virus into one side of the thalamus. This figure was based on Morita *et al.* [13].

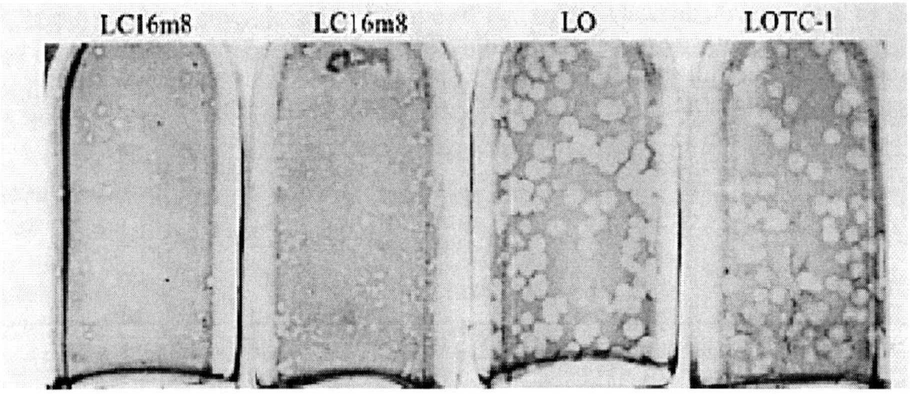

Figure 2. Plaques of LC16m8, LO and LOTC-1 formed in the monolayer of rabbit kidney cells. LOTC-1 was made by inserting a DNA fragment of LO containing the B5R gene into LC16m8. *TK* and *HA* indicate thymidine kinase and hemagglutinin genes, respectively. These data are cited from Takahashi-Nishimaki *et al.* [16].

Table 2. Properties of LO and its attenuated variants, LC16mO, LC16m8 and LOTC.

	LO	**LC16mO**	**LC16m8**	**LOTC**
Temperature sensitivity (ceiling temperature, °C)	>45	40.8	40.8	40.8
Plaque size (rabbit kidney cells)	Large	Medium	Small	Large
Pock size in chorioallantoic membrane	Large	Medium	Small	Large
Viral growth in cultured cells				
Rabbit kidney cells	+++	+++	+++	+++
Vero cells	+++	++	+	+++
Viral growth in the brain				
Rabbit	++	+	+ or -	+
Monkey	+++	+	+	n.d.
Dermal pathology (rabbit)	+++	++	+	n.d.
Antibody response (rabbit)	++	+++	++	n.d.
B5R (*ps/hr*) gene	Intact	Intact	Frame shift mutation	Intact

This table is cited from the reference 27 with permission from Elsevier Ltd.

Clinical Studies

Strain mO was vaccinated into 829 Japanese people, and m8 was vaccinated into more than 50,000 children in Japan during 1974-1975, without inducing serious side effects.

In 1975, the National Smallpox Vaccination Research Committee summarized in a report the results of clinical studies on LC16, mO and m8 strains [4]. 1) No problematic side effects were observed among 50,000 children inoculated with m8 in 1974. 2) Among 10,578 people who were vaccinated with the m8 strain during 1973 and 1974, and were examined in detail for clinical symptoms, the rate of take was 95.1%, the average incidence of fever was 7.7%, the incidence of lymph node swelling was 12-13%, and the mean diameter of induration was 6.1 mm (Table 3). 3) The ability by mO and m8 to induce neutralizing antibodies and hemagglutinin-inhibiting titers was comparable to those by the conventional vaccine strains (CV-1 and Lister strains). 4) The protective immunity of LC16 assessed by the suppression of the primary response of pus blister formation was comparable to the conventional vaccine (Ikeda strain) (18.8% for m8 and 23.0% for the conventional vaccine at 12 months after inoculation). 5) Brain waves were examined by an electroencephalograph of 56 subjects inoculated with m8 at 14 days after inoculation: no subject showed abnormal signs. Even the mO strain showing severer dermal pathogenesis than m8 was attenuated in terms of fever response by inoculating 829 subjects.

Based on these studies, m8 was licensed in 1975 in Japan as a second-generation smallpox vaccine. It is prepared by using a cell culture system for the first time in the world because the use of cell culture made possible the cloning of viruses so as to avoid the problem of adventitious viruses and contamination by microbes, such as bacteria. The vaccine was stockpiled after lyophilization. Strain m8 has enough immunogenicity for protection in animal experiments, as described in detail later. Thus the m8 strain had already fulfilled almost all the criteria as a modern, second-generation vaccine in the early 1970s. However, since the success of smallpox eradication by the WHO program, the m8 strain is unlikely to show its effectiveness against endemic smallpox.

Table 3. Reaction of various VV strains.

Strains	Year of investigation	Number of cases	Rate of take (%)	Mean diameter of induration (mm)	Rate of fever (%)
Ikeda	1968 ~ 70	1,506	99.1	18.2	25.0
Ecuador	1969 ~ 70	1,846	67.5	17.4	21.3
LO	1968 ~ 71	3,662	93.7	15.3	26.6
CV-1	1971 ~ 73	22,976	92.4	16.8	8.5
LC16mO	1973 ~ 74	829	94.8	14.5	12.1
LC16m8	1973 ~ 74	10,578	95.1	6.1	7.7

Cited from the reference by Yamaguchi et al. [4]

The *B5R* Gene

A series of genetic studies of these temperature-sensitive strains by Sugimoto, Takahashi-Nishimaki *et al.* [14-16] identified a viral gene responsible for plaque size and the restriction of the host range.

First, analyses of genome DNAs by digestion with restriction endonucleases and cross-hybridization of the digested fragments showed that mO and m8 have a new *Xho*I site, in the *HinD*III D fragment, in addition to the 14 *Xho*I sites of LO [14] (Figure 3). According to preliminary studies at that time, the LC16 variants and LO showed no marked difference in the genome structures, except for the new *Xho*I site and their terminal fragments that were not identified in LO owing to their heterogeneity.

From the result suggesting a possibility that the *Hind*III D fragment contains a gene responsible for attenuation of mO and m8, VV variants were prepared by introducing part of the *Hind*III D fragment of the DNA of the parental LO strain into the m8 strain [16]. Five LO virus variants (LOTC-1 to LOTC-5) forming much larger plaques than LC16m8 were cloned (Figure 2). In LOTC-2, LOTC-4 and LOTC-5, the introduction of at least part of *Hind*III D of LO into the corresponding *Hind*III D region of the LC16m8 genome was confirmed as judged by the disappearance of the *Xho*I site, whereas variants LOTC-1 and LOTC-3 retained the *Xho*I site. This result suggested that the putative gene responsible for plaque size is near, but does not necessarily contain, the unique *Xho*I site. The biological characteristics of all LOTC variants were similar. Their plaque size and pock size were similar to those of LO, but they were resembled LC16m8 for temperature sensitivity and neurovirulence. These results strongly suggested that part of the *Hind*III D fragment participates in determining biological characteristics affecting plaque size and pock size, but has little influence on temperature sensitivity.

The mapping and identification of the putative gene in the m8, mO and LO genomes responsible for plaque size was done by introducing various subfragments of the *Hind*III D fragment of LO DNA into the gene of the m8 strain and by screening the resultant virus variants for their ability to form large plaques [15]. The results indicated that an open reading frame in LO *Hind*III D fragment was responsible for plaque size. The gene also participates in the host range, that is, the ability to grow in Vero cells, and was named as *plaque size/host range* (*ps/hr*) gene. Because the complete sequencing of the *ps/hr* gene confirmed that the unique *Xho*I site was not contained in this gene, the choice of the *Xho*I site as a marker to pursue the *ps/hr* gene was accidental and lucky.

The *ps/hr* gene encodes a polypeptide of 317 amino acids related to the regulators of complement activation gene family of mammals. Thus, the genetic analysis described here provided direct evidence for a previously unrecognized function of the regulators of complement activation-related proteins encoded by a virus [17], called *B5R* gene. The name "*B5R* gene" is used instead of "*ps/hr* gene", because "*B5R*" is now widely used. The m8 strain has a single nucleotide deletion that creates a termination codon at amino acid position 93 in the *B5R* evelope gene, resulting in a truncation of the B5R protein.

After the function of *B5R* gene was discovered, study of the biological function of the B5R protein has been active by other investigators. Their results also supported our early finding for m8: destruction of the *B5R* gene in VV contributes to small plaques and attenuation (review, [18]). VV produces four different types of virion: intracellular mature virus, intracellular enveloped virus, cell-associated enveloped virus and extracellular enveloped virus. Extracelluar enveloped viruses are free virions released from infected cells and may cause long-range dissemination of infection, although they are less than 1% of the virus population, of which the majority are intracellular mature viruses. Seven virus-encoded proteins have been identified that are components of intracellular enveloped viruses; five of them are in cell-associated enveloped viruses or extracellular enveloped viruses. The B5R protein is contained in the envelope of all these three enveloped viruses. These proteins, including B5R, have a role in virus morphogenesis and dissemination, and as targets for neutralizing antibodies.

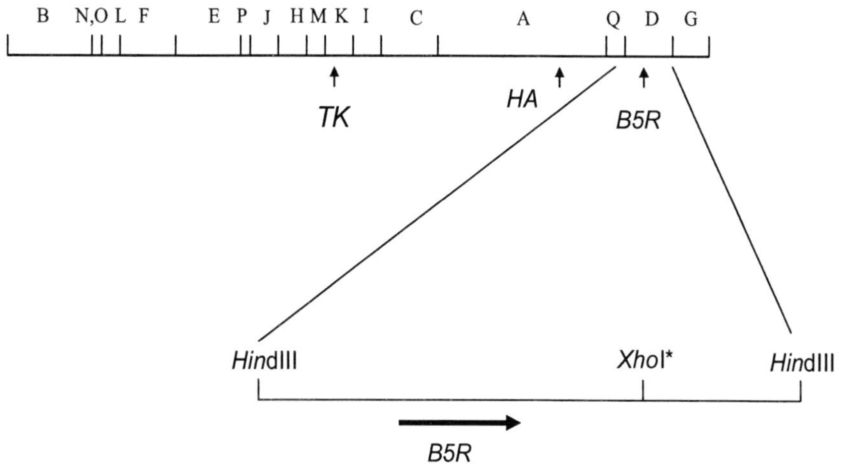

Figure 3. Physical map of the location of the *B5R* gene in *Hin*dIII restriction fragments of DNA from LO, mO and m8 strains. *The unique *Xho*I site in mO and m8 DNA, but not in LO DNA. This figure is cited from Sugimoto *et al.* [45]

Full Genome Sequence of m8 and mO

Recently, Morikawa *et al.* [19] determined the full genome sequences of the m8, parental mO and grand parental LO strains, and analyzed their evolutionary relationships. The results indicated that m8 is a progeny of LO that preserves almost all the open reading frames of VV, except for the disrupted *B5R* gene.

A detailed comparison of m8, mO and LO shows that because m8 and mO are cloned viruses, their genome sequences are easily compared. Their gene structures and organizations are almost the same: only six point mutations are between m8 and mO. Three point mutations are in noncoding regions, and a single-amino-acid substitution is in four open reading frames. The most remarkable change is a deletion of G in the *B5R* gene, resulting in a termination codon and truncation of the B5R protein. LO has no history of virus cloning, and an analysis of nucleotide polymorphisms showed that LO consists of a huge divergent virus population but likely contains ancestors of mO.

The above genetic studies suggest the following points of the evolution of m8 from LO. During the first stage of selection of LC16 from LO, temperature-sensitive clones were selected from the original divergent virus population. The mO strain was cloned from an LC16 virus population that had a medium-size pock and temperature-sensitivity. The m8 strain with a frame-shift mutation of the *B5R* gene was finally cloned from an mO virus population that had a small pock. Only four missense mutations were among the more than 280 open reading frames of m8 during evolution from the parental mO strain. A major change is a truncating mutation of the B5R gene. Thus, inactivation of the B5R protein may mainly contribute to the attenuation of the m8 strain, at least partly due to a decreased ability in propagation. However, the remaining three missense mutations may also contribute to the attenuation of m8, including low neurovirulence. The genes responsible for temperature sensitivity of LC16 derived from LO remain to be clarified.

Immunological and Pathological Properties

Early studies of m8 showed highly attenuated properties of m8 in humans and strong immunogenicity in rabbits and monkeys comparable to that of LO used as a vaccine strain. However, the fact that m8 has the truncated B5R protein and that B5R is the only neutralizing antibody-inducing antigen of extracellular enveloped viruses so far identified [20] raises the concern that m8 might have a defect in the protective immunity against smallpox.

Morikawa *et al.* [19] studied the protective efficacies of m8 against a lethal challenge with a highly pathogenic VV, WR strain, in mice. Vaccinated BALB/c mice were examined for survival rate and loss of body weight after intranasal inoculation with 10^6 PFU of WR corresponding to a 10-times dose of LD_{50} for these mice. Non-immunized mice that received phosphate-buffered saline developed clinical symptoms, loss of body weight and died within nine days after the WR challenge. In contrast, none of the mice vaccinated with m8, mO or LO died. Notably, no marked difference was in body weight between low-dose (10^5 PFU) and high-dose (10^7 PFU) vaccination groups, nor between the m8, mO and LO vaccination groups. Kidokoro *et al.* [21] also reported protective immunity of m8 against a challenge with a lethal dose of WR viruses in BALB/c mice.

The m8 vaccine is unlikely to be used for a large population in endemic areas because smallpox was almost eradicated when the vaccine was developed. Today, no vaccines can be tested for protective efficacy against infection of humans with the variola virus. Thus, the results of the protection of mice against the pathogenic WR strain provide additional evidence to evaluate protective immunization.

Among the more than 280 open reading frames in the VV genome, some of the open reading frames may code proteins that function as protective antigens. The only protein that is deleted in m8 is B5R. Although the B5R protein induces neutralizing antibodies [20], many other proteins, including envelope proteins such as A33R, A34R and A56R, probably have also an important role in protection. Not only neutralizing antibodies, but also cell-mediated immunity, is suggested to participate in protection against orthopoxviruses. Although purified B5R protein induces B5R-specific neutralizing antibodies in rabbits, such antibodies are not induced by immunization by natural infection or by VV virions, meaning that the B5R protein is a "hidden antigen" [20]. All these facts further support the idea that the absence of normal B5R protein in m8 is never a serious problem for m8 to function as a vaccine.

The Problem of Revertant

Kidokoro *et al.* [21] found that m8 reverted spontaneously to a large plaque-forming clone. The *B5R* gene was responsible for the revertant formation: a single nucleotide insertion just upstream of the m8 *B5R* mutation site restored the open reading frame to the parental mO phenotype after ten or more virus passages. To avoid this problem, Kidokoro *et al.* [21] constructed a genetically stable and fully effective vaccine strain by deleting the *B5R* gene from m8. This virus strain may be a future candidate vaccine to replace m8.

For the moment, however, the management of vaccine preparation can cope effectively with this problem and the revertant content included in the vaccine has been controlled at low levels.

As a Vector for Recombinant Vaccine

The attenuated VV is useful not only as a smallpox vaccine, but also as a vector to express foreign proteins. The recombinant vaccinia virus (RVV) vaccine is expected to be a promising live vaccine against pathogens with foreign antigens because it can induce both strong humoral and cell-mediated immunity [22,23]. Successful eradication of rabies virus in a large-scale campaign of fox vaccination in Belgium paved the way for practical application of an RVV vaccine [24,25]. The rabies RVV has been used in the field as an oral vaccine against rabies of foxes in Europe and of raccoons and striped skunks in North America.

Because m8 and related strains are highly immunogenic but are low in neurovirulence, we developed several kinds of RVVs expressing foreign pathogenic antigens by using m8 and mO as a vector. One early study [26] showed that RVVs preserve the attenuated properties and heat-stability of VV vectors. Because RVVs using m8 as a vector showed much less replicating activity in cultured cells than RVVs using mO, mO was later used as the vector of RVVs. The results of RVVs using m8 and mO strains as a vector were summarize in our review [27]. Here in this review, we focussed on experiments that used animals inoculated with RVVs expressing antigens of rinderpest virus (RPV) and bovine leukemia virus (BLV), whose efficacies were tested in large animals.

Rinderpest Recombinant Vaccine

RPV, of the genus *Morbillivirus* in the family *Paramyxoviridae*, is an etiological agent of rinderpest. It causes acute and often fatal disease in cattle and large ruminants and serious economic loss in domestic live stock enterprises. Sporadic outbreaks still occur in parts of Africa, the Middle East and south Asia.

RVVs expressing the hemagglutinin gene from RPV in mO were constructed [28,29]. The following results were obtained for the properties of RVV and for experiments on cattle inoculated with RVV and then a challenge with lethal doses of RPV [30-32].

1) *Heat stability:* the RVVs preserved heat stability comparable to their VV vector mO.
2) *Safety:* Four cattle were inoculated with 10^8 p.f.u. of the vaccine, and two contact control cattle were not vaccinated. Two other cattle kept in a separate room were not inoculated. The vaccinated cattle showed neither febrile responses nor skin lesions at the site of inoculation. No virus was isolated from peripheral blood leukocytes, or from nasal, ocular and rectal swabs. The two contact control cattle did not develop anti-VV or anti-rinderpest virus antibodies. These findings indicate that recombinant vaccine has little risk of being released into the environment from inoculated cattle or of spreading to animals in contact with inoculated cattle. .
3) *Protective effect:* Three weeks after the inoculation, cattle were subcutaneously challenged with 10^4 $TCID_{50}$ of a virulent Saudi 1/81 strain of RPV. All vaccinated cattle survived against the lethal challenge without clinical symptoms. In contrast, the four control cattle, including the two contact control cattle, developed a fatal typical acute rinderpest disease .
4) *Minimal dose:* To find a minimum immunizing dose, 12 cattle were used: each of five pairs of cattle was inoculated with 10^8, 10^7, 10^6, 10^5 or 10^4 p.f.u. of vaccine and two cattle were not inoculated. All RVV-inoculated cattle survived, except for one cattle inoculated with 10^5 p.f.u. RVV, but the two control cattle developed a fatal disease [30]. These results showed that a dose as low as 10^4 p.f.u. of vaccine is effective.
5) *Duration of immunity:* To examine the duration of protective immunity, the 12 cattle with 10^8 p.f.u. of vaccine were challenged every year for up to three years. An additional vaccination as a booster was not given. All vaccinated cattle survived the lethal challenge without clinical signs in a one-year trial [31], and all vaccinated cattle, except one, survived the lethal challenge in two-year and three-year trials [32]. One vaccinated cattle inoculated with RVV in the two-year trial showed delayed transient fever and a very mild stomatitis, and died by a fatal black-leg infection. These results showed that the rinderpest RVV protected cattle against the lethal challenge of RPV at least for three years. VV replicates locally in squamous epidermal cells without causing viremia. Surprisingly, such a limited local growth of the virus provided a long-lasting immunity against systemic infection with RPV. Because we used the highly virulent Saudi 1/81 strain of RPV that can kill cattle within three days as the challenge virus in these experiments, we consider this RVV to sufficiently protect cattle in the field for a long period.
6) *Role of humoral and cell-mediated immunity:* The RVV achieved the protection by extremely low levels of neutralizing antibodies compared with the present live vaccine, RBOK strain [33]. The antibody titers sharply increased after the challenge [30-32]. These results suggest that cell-mediated immunity also participates in protection by the RVV.

Bovine Leukemia Virus (BLV) Recombinant Vaccine

BLV, of the Retroviridae, causes malignancy of the B-cell lineage termed enzootic bovine leukemia in cattle. It shares biological and pathological properties with human T-cell leukemia virus types I and II. It induces high incidences of leukemia in sheep within a relatively short period, and so sheep are a good animal model for research on BLV.

RVVs expressing BLV envelope glycoprotein gp60 were constructed using mO as a vector [34]. Nine sheep were inoculated with 10^8 p.f.u. of RVV vaccine. At 20 weeks after vaccination, the nine vaccinated sheep and ten unvaccinated sheep were challenged with 10^5 BLV-infected bovine lymphocytes [35]. BLV titers of vaccinated sheep were markedly suppressed compared with titers of control sheep, although BLV was not completely eliminated from the RVV-vaccinated sheep. No specific antibody against BLV was induced by RVV vaccination, but the antibodies increased sharply after the challenge. However, the vaccination induced strong delayed hypersensitivity responses [35,36]. These results suggested that cell-mediated immunity participates in suppression of BLV replication.

We examined the efficacy of RVV as a prophylactic vaccine. Five sheep were challenged with BLV-infected bovine lymphocytes, and then were inoculated with 10^8 p.f.u. RVV. BLV titers of vaccinated animals were suppressed after vaccination, but titers of control animals increased [35]. BLV titers were inversely correlated with specific lymphocyte proliferation response, but they were not correlated with anti-BLV antibodies [37]. These results strongly suggested that cell-mediated immunity, not humoral immunity, is responsible for suppression of BLV titers, and would provide an important insight for vaccine development against retroviruses in humans, including human immunodeficiency virus [38].

Results of several studies of RVVs, including RVVs expressing RPV or BLV antigens, showed that an RVV using mO as a vector preserves its attenuated properties [26,27,30,35,39]. The results of RVV expressing the BLV antigen strongly suggested the importance of cell-mediated immunity in protection [35,37]. These results are consistent with the view that T-cell immunity induced by vaccination with VV has an important role in protection against smallpox [40-43].

Earlier studies used mO, not m8, as a vector of RVVs because RVVs using m8 become less active in replication than the parental virus m8 [26]. However, the use of a strong promoter, such as developed by Kidokoro *et at.* [44], may make it possible to use m8 as a vector for RVV.

CONCLUSION

Although m8 vaccine is unlikely to be used for a large population in endemic areas because smallpox was almost eradicated when it was developed, early studies of humans and large animals, including monkey, support the view that m8 vaccine has a very low neurovirulence and high immunogenicity. Recent studies of mice showed that m8 induces a hundred per cent protection against the highly pathogenic vaccinia WR virus. The full-genome sequences of m8, mO and parental Lister (LO) strains used for smallpox eradication showed that m8 preserves all open reading frames of VV, except for the disrupted envelope gene *B5R*, supporting the idea that m8 has enough molecules responsible for protection, comparable to those of LO. All these facts together support the view that m8 is a safe and effective vaccine against smallpox by variola virus infection.

ACKNOWLEDGEMENTS

We thank the following researchers who cooperated with us: the late Dr. Michio Morita (then the Chiba Serum Institute), Dr. Atsushi Yasuda (Nippon Zeon Co., Ltd.), Dr. Tadashi Maruyama (Japan Agency for Marine-Earth Science and Technology), Dr. Fukumi Takahashi-Nishimaki (then Tonen Co. Ltd.), Dr. Yoji Ikawa (Ikawa Laboratory, Riken), Professor Emeritus Kazuya Yamanouchi (the University of Tokyo), Dr. Thomas Barrett (Institute for Animal Health, UK), Professor Kosuke Okada (Iwate University). We also thank Drs. Asato Kojima and Minoru Kidokoro (National Institute of Infectious Diseases), and Dr. Isao Arita (Chairman of the Agency for Cooperation in International Health) for sending us invaluable information and comments.

REFERENCES

[1] World Health Organization. Declaration of global eradication of smallpox. *Weekly Epidemiological Record 148*, 1980.

[2] Arita, I. Smallpox vaccine and its stockpile in 2005. Lancet Infect Dis, 5: 647-652, 2005.

[3] Lane, J. M., Ruben, F. L., Neff, J. M., and Millar, J. D. Complications of smallpox vaccination, 1968. *N Engl J Med, 281*: 1201-1208, 1969.

[4] Yamaguchi, M., Kimura, M., and Hirayama, M. Report of the National Smallpox Vaccination Research Committee: study of side effects, complications and their treatment. (in Japanese). *Clinical Virology, 3*: 269-278, 1975.

[5] Weltzin, R., Liu, J., Pugachev, K. V., Myers, G. A., Coughlin, B., Blum, P. S., Nichols, R., Johnson, C., Cruz, J., Kennedy, J. S., Ennis, F. A., and Monath, T. P. Clonal vaccinia virus grown in cell culture as a new smallpox vaccine. *Nat Med, 9*: 1125-1130, 2003.

[6] Earl, P. L., Americo, J. L., Wyatt, L. S., Eller, L. A., Whitbeck, J. C., Cohen, G. H., Eisenberg, R. J., Hartmann, C. J., Jackson, D. L., Kulesh, D. A., Martinez, M. J., Miller, D. M., Mucker, E. M., Shamblin, J. D., Zwiers, S. H., Huggins, J. W., Jahrling, P. B., and Moss, B. Immunogenicity of a highly attenuated MVA smallpox vaccine and protection against monkeypox. *Nature, 428*: 182-185, 2004.

[7] Belyakov, I. M., Earl, P., Dzutsev, A., Kuznetsov, V. A., Lemon, M., Wyatt, L. S., Snyder, J. T., Ahlers, J. D., Franchini, G., Moss, B., and Berzofsky, J. A. Shared modes of protection against poxvirus infection by attenuated and conventional smallpox vaccine viruses. *Proc Natl Acad Sci U S A, 100*: 9458-9463, 2003.

[8] Chen, R. T. and Lane, J. M. Myocarditis: the unexpected return of smallpox vaccine adverse events. *Lancet, 362*: 1345-1346, 2003.

[9] Enserink, M. Biodefense. Smallpox vaccines: looking beyond the next generation. *Science, 304*: 809, 2004.

[10] Hashizume, S., Morita, M., Yoshikawa, H., Suzuki, S., Arita, T., Komatsu, T., Amano, H., and Tachibana, Y. Intracerebral inoculation of monkeys with several vaccinia strains.

Internal symposium on smallpox vaccines. *Bilthoven. Symp. Ser. Immunobiol. Stand., 19*: 325-331, 1973.

[11]Hashizume, S., Yoshikawa, H., Morita, M., and Suzuki, K. (eds.) *Properties of attenuated mutant of vaccinia virus, LC16m8, derived from Lister strain.*, p. pp421-428. Amsterdam, The Netherland: Elsevier Science Publishing Co., 1985.

[12]Kurata, T. Coping with bioterrorism (in Japanese). 2002. http://idsc.nih.go.jp/training/14kanri/kurata.htm

[13]Morita, M., Aoyama, Y., Arita, M., Amona, H., Yoshizawa, H., Hashizume, S., Komatsu, T., and Tagaya, I. Comparative studies of several vaccinia virus strains by intrathalamic inoculation into cynomolgus monkeys. *Arch Virol, 53*: 197-208, 1977.

[14]Sugimoto, M., Yasuda, A., Miki, K., Morita, M., Suzuki, K., Uchida, N., and Hashizume, S. Gene structures of low-neurovirulent vaccinia virus LC16m0, LC16m8, and their Lister original (LO) strains. *Microbiol Immunol, 29*: 421-428, 1985.

[15]Takahashi-Nishimaki, F., Funahashi, S., Miki, K., Hashizume, S., and Sugimoto, M. Regulation of plaque size and host range by a vaccinia virus gene related to complement system proteins. *Virology, 181*: 158-164., 1991.

[16]Takahashi-Nishimaki, F., Suzuki, K., Morita, M., Maruyama, T., Miki, K., Hashizume, S., and Sugimoto, M. Genetic analysis of vaccinia virus Lister strain and its attenuated mutant LC16m8: production of intermediate variants by homologous recombination. *J Gen Virol, 68*: 2705-2710., 1987.

[17]Goebel, S. J., Johnson, G. P., Perkus, M. E., Davis, S. W., Winslow, J. P., and Paoletti, E. The complete DNA sequence of vaccinia virus. *Virology, 179*: 247-266, 517-263., 1990.

[18]Smith, G. L., Vanderplasschen, A., and Law, M. The formation and function of extracellular enveloped vaccinia virus. *J Gen Virol, 83*: 2915-2931, 2002.

[19]Morikawa, S., Sakiyama, T., Hasegawa, H., Saijo, M., Maeda, A., Kurane, I., Maeno, G., Kimura, J., Hirama, C., Yoshida, T., Asahi-Ozaki, Y., Sata, T., Kurata, T., and Kojima, A. An attenuated LC16m8 smallpox vaccine: analysis of full-genome sequence and induction of immune protection. *J Virol, 79*: 11873-11891, 2005.

[20]Galmiche, M. C., Goenaga, J., Wittek, R., and Rindisbacher, L. Neutralizing and protective antibodies directed against vaccinia virus envelope antigens. *Virology, 254*: 71-80, 1999.

[21]Kidokoro, M., Tashiro, M., and Shida, H. Genetically stable and fully effective smallpox vaccine strain constructed from highly attenuated vaccinia LC16m8. *Proc Natl Acad Sci U S A, 102*: 4152-4157, 2005.

[22]Panicali, D. and Paoletti, E. Construction of poxviruses as cloning vectors: insertion of the thymidine kinase gene from herpes simplex virus into the DNA of infectious vaccinia virus. *Proc Natl Acad Sci U S A, 79*: 4927-4931., 1982.

[23]Mackett, M., Smith, G. L., and Moss, B. Vaccinia virus: a selectable eukaryotic cloning and expression vector. *Proc Natl Acad Sci U S A, 79*: 7415-7419, 1982.

[24]Brochier, B., Kieny, M. P., Costy, F., Coppens, P., Bauduin, B., Lecocq, J. P., Languet, B., Chappuis, G., Desmettre, P., Afiademanyo, K., and et al. Large-scale eradication of rabies using recombinant vaccinia-rabies vaccine. *Nature, 354*: 520-522, 1991.

[25]Brochier, B., Aubert, M. F., Pastoret, P. P., Masson, E., Schon, J., Lombard, M., Chappuis, G., Languet, B., and Desmettre, P. Field use of a vaccinia-rabies recombinant vaccine for the control of sylvatic rabies in Europe and North America. *Rev Sci Tech, 15*: 947-970, 1996.

[26]Morita, M., Suzuki, K., Yasuda, A., Kojima, A., Sugimoto, M., Watanabe, K., Kobayashi, H., Kajima, K., and Hashizume, S. Recombinant vaccinia virus LC16m0 or LC16m8 that expresses hepatitis B surface antigen while preserving the attenuation of the parental virus strain. *Vaccine, 5*: 65-70., 1987.

[27]Sugimoto, M. and Yamanouchi, K. Characteristics of an attenuated vaccinia virus strain, LC16m0, and its recombinant virus vaccines. *Vaccine, 12*: 675-681., 1994.

[28]Tsukiyama, K., Yoshikawa, Y., Kamata, H., Imaoka, K., Asano, K., Funahashi, S., Maruyama, T., Shida, H., Sugimoto, M., and Yamanouchi, K. Development of heat-stable recombinant rinderpest vaccine. *Arch Virol, 107*: 225-235, 1989.

[29]Asano, K., Tsukiyama, K., Shibata, S., Yamaguchi, K., Momoki, T., Maruyama, T., Kohara, M., Miki, K., Sugimoto, M., Yoshikawa, Y., and et al. Immunological and virological characterization of improved construction of recombinant vaccinia virus expressing rinderpest virus hemagglutinin. *Arch Virol, 116*: 81-90, 1991.

[30]Yamanouchi, K., Inui, K., Sugimoto, M., Asano, K., Nishimaki, F., Kitching, R. P., Takamatsu, H., and Barrett, T. Immunisation of cattle with a recombinant vaccinia vector expressing the haemagglutinin gene of rinderpest virus. *Vet Rec, 132*: 152-156., 1993.

[31]Inui, K., Barrett, T., Kitching, R. P., and Yamanouchi, K. Long-term immunity in cattle vaccinated with a recombinant rinderpest vaccine. *Vet Rec, 137*: 669-670., 1995.

[32]Ohishi, K., Inui, K., Barrett, T., and Yamanouchi, K. Long-term protective immunity to rinderpest in cattle following a single vaccination with a recombinant vaccinia virus expressing the virus haemagglutinin protein. *J Gen Virol, 81*: 1439-1446., 2000.

[33]Plowright, W. and Ferris, R. Studies with rinderpest virus in tissue culture virus as a vaccine for cattle. *Res. Vet. Sci., 3*: 172-182, 1962.

[34]Ohishi, K., Suzuki, H., Maruyama, T., Yamamoto, T., Funahashi, S., Miki, K., Ikawa, Y., and Sugimoto, M. Induction of neutralizing antibodies against bovine leukosis virus in rabbits by vaccination with recombinant vaccinia virus expressing bovine leukosis virus envelope glycoprotein. *Am J Vet Res, 51*: 1170-1173., 1990.

[35]Ohishi, K., Suzuki, H., Yamamoto, T., Maruyama, T., Miki, K., Ikawa, Y., Numakunai, S., Okada, K., Ohshima, K., and Sugimoto, M. Protective immunity against bovine leukaemia virus (BLV) induced in carrier sheep by inoculation with a vaccinia virus-BLV env recombinant: association with cell-mediated immunity. *J Gen Virol, 72*: 1887-1892., 1991.

[36]Okada, K., Ikeyama, S., Ohishi, K., Suzuki, H., Sugimoto, M., Numakunai, S., Chiba, T., Murakami, K., Davis, W. C., Ohshima, K., and et al. Involvement of CD8+ T cells in delayed-type hypersensitivity responses against bovine leukemia virus (BLV) induced in sheep vaccinated with recombinant vaccinia virus expressing BLV envelope glycoprotein. *Vet Pathol, 30*: 104-110., 1993.

[37]Ohishi, K., Suzuki, H., Yasutomi, Y., Onuma, M., Okada, K., Numakunai, S., Ohshima, K., Ikawa, Y., and Sugimoto, M. Augmentation of bovine leukemia virus (BLV)-specific lymphocyte proliferation responses in ruminants by inoculation with BLV env-recombinant vaccinia virus: their role in the suppression of BLV replication. *Microbiol Immunol, 36*: 1317-1323, 1992.

[38]Sugimoto, M., Ohishi, K., and Ikawa, Y. Cell-mediated immunity to BLV replication: analogous to HIV immunopathogenesis? *Immunol Today, 14*: 190-191., 1993.

[39]Shida, H., Hinuma, Y., Hatanaka, M., Morita, M., Kidokoro, M., Suzuki, K., Maruyama, T., Takahashi-Nishimaki, F., Sugimoto, M., Kitamura, R., and et al. Effects and

virulences of recombinant vaccinia viruses derived from attenuated strains that express the human T-cell leukemia virus type I envelope gene. *J Virol, 62*: 4474-4480., 1988.

[40]Drexler, I., Staib, C., Kastenmuller, W., Stevanovic, S., Schmidt, B., Lemonnier, F. A., Rammensee, H. G., Busch, D. H., Bernhard, H., Erfle, V., and Sutter, G. Identification of vaccinia virus epitope-specific HLA-A*0201-restricted T cells and comparative analysis of smallpox vaccines. *Proc Natl Acad Sci U S A, 100*: 217-222, 2003.

[41]Terajima, M., Cruz, J., Raines, G., Kilpatrick, E. D., Kennedy, J. S., Rothman, A. L., and Ennis, F. A. Quantitation of CD8+ T cell responses to newly identified HLA-A*0201-restricted T cell epitopes conserved among vaccinia and variola (smallpox) viruses. *J Exp Med, 197*: 927-932, 2003.

[42]Combadiere, B., Boissonnas, A., Carcelain, G., Lefranc, E., Samri, A., Bricaire, F., Debre, P., and Autran, B. Distinct time effects of vaccination on long-term proliferative and IFN-gamma-producing T cell memory to smallpox in humans. *J Exp Med, 199*: 1585-1593, 2004.

[43]Kennedy, J. S., Frey, S. E., Yan, L., Rothman, A. L., Cruz, J., Newman, F. K., Orphin, L., Belshe, R. B., and Ennis, F. A. Induction of human T cell-mediated immune responses after primary and secondary smallpox vaccination. *J Infect Dis, 190*: 1286-1294, 2004.

[44]Kidokoro, M., Aoki, A., Horiuchi, K., and Shida, H. Large-scale preparation of biologically active measles virus haemagglutinin expressed by attenuated vaccinia virus vectors. *Microbes Infect, 4*: 1035-1044., 2002.

[45]Sugimoto, M., Ohishi, K., Kidokoro, M., and Hashizume, S. A new role of domestic smallpox vaccine (in Japanese). *Protein, Nucleic Acid and Enzyme, 48*: 1693-1701, 2003.

In: Focus on Bioterrorism
Editor: George T. Foster, pp. 17-23
ISBN 1-60021-185-2

Chapter 2

SMALLPOX VACCINE STOCKPILE AND VACCINATION POLICY

Judith A. Johnson

ABSTRACT

On June 20, 2002, an advisory panel to the U.S. Centers for Disease Control and Prevention (CDC) voted against mass smallpox vaccination and instead recommended that 10-20 thousand first responders receive the vaccine. The recommendations are currently under review by CDC and the Department of Health and Human Services (HHS). Congress addresses control of the vaccine stockpile in the Homeland Security legislation (H.R. 5005, S. 2452). Although there is little uneasiness over the safety of the only known samples of variola virus, held by the CDC in Atlanta and a laboratory near Novosibirsk in Russia, there is some concern that samples of the virus may have been acquired by terrorists or rogue governments, particularly because of the unrest that occurred during the break up of the Soviet Union. The Public Health Security and Bioterrorism Preparedness and Response Act of 2002 (P.L. 107-188), signed by the President on June 12, 2002, directs the Secretary of HHS to ensure that there is enough smallpox vaccine in the Strategic National Stockpile to meet health security needs and authorizes $509 million for FY2002 and such sums as may be necessary through FY2006 for this purpose. H.R. 5005, the Homeland Security Act of 2002 as passed by the House, would amend P.L. 107-188 by moving authority for the stockpile to the Department of Homeland Security (DHS); HHS would continue to manage the stockpile and determine its contents. S. 2452, the National Homeland Security and Combating Terrorism Act of 2002, as reported by the Governmental Affairs Committee on July 25, 2002, directs DHS to consult with CDC in the administration of the stockpile by CDC. An emergency supplemental appropriation (P.L. 107-117), signed by the President on January 10, 2002, provided $512 million for the purchase of smallpox vaccine by HHS.

On June 20, 2002, an advisory panel to the U.S. Centers for Disease Control and Prevention (CDC) voted against mass smallpox vaccination and instead recommended that 10-20 thousand first responders receive the vaccine. The recommendations are currently under review by CDC and the Department of Health and Human Services (HHS). Congress addresses control of the vaccine stockpile in the Homeland Security legislation (H.R. 5005, S. 2452).

The terrorist attacks of the fall of 2001 resulted in increased attention to preparedness for other such possible events, one being the deliberate release of variola virus, the virus which causes smallpox. Although there is little uneasiness over the safety of the only known samples of variola virus, held by the CDC in Atlanta and a laboratory near Novosibirsk in Russia, there is some concern that samples of the virus may have been acquired by terrorists or rogue governments, particularly because of the unrest that occurred during the break up of the Soviet Union.

Smallpox is transmitted via person-to-person contact or inhalation of saliva droplets in the breath of an infected person. The incubation period of 12-14 days (range 7-17 days) is followed by the sudden onset of flu-like symptoms. After 2-3 days the distinctive rash begins to appear, and it is this stage, which lasts for 7-10 days, that is most infectious. As scabs begin to form infectivity rapidly declines. About 65% to 80% of survivors have deep scars or pockmarks that tend to occur on the face. After a worldwide effort of organized vaccination, smallpox was declared eradicated by the World Health Assembly in May 1980.[1] Smallpox vaccination is achieved via inoculation using live vaccinia virus, a virus linked with cowpox. Vaccination is highly effective and can prevent smallpox even when given 2 to 3 days after exposure to variola. The last recorded natural case of smallpox occurred in Somalia in 1977; a fatal laboratory-acquired case occurred in the United Kingdom in 1978. The last U.S. case was in 1949. The U.S. Public Health Service recommended in 1971 that routine smallpox vaccination be discontinued and the U.S. program stopped in 1972.

An unfortunate side effect of the eradication of smallpox, widely acclaimed as a major public health triumph, is that the U.S. population and the rest of the world are vulnerable to biologic weapon attack using variola virus. In past natural outbreaks, smallpox killed 30% of those who had no immunity. Approximately 119 million U.S. residents born after 1972 lack immunity. The immunity of the 157 million U.S. residents born before 1972 has declined and their level of disease protection is uncertain.

Because of these facts, CDC and HHS asked the Advisory Committee on Immunization Practices (ACIP) to consider if changes were needed in the June 2001 recommendations on smallpox vaccination.[2] Under the June 2001 recommendations, only laboratory workers who directly handle smallpox-related virus are vaccinated. After several public meetings, draft recommendations on smallpox vaccination were approved by ACIP on June 20, 2002. ACIP did not recommend smallpox vaccination of the general public because "the risk of an attack was assessed to be low"...[and] "the potential benefits of vaccination do not outweigh the risks of vaccine complications." Instead ACIP recommended that 10 to 20 thousand first responders receive the vaccine. These individuals would act in teams investigating initial smallpox cases and vaccinating the patient's contacts. Under the ACIP plan, healthcare personnel at a limited number of hospitals involved in caring for smallpox patients would also be vaccinated.

---

1 The World Health Assembly determines policy and budget for the World Health Organization.

2 ACIP is a 15 member group of medical and public health experts that provides advice to HHS and CDC on vaccine use and policy for effective disease control in the civilian population.

The ACIP draft recommendations are now being considered by CDC and HHS.[3] Media reports of July 7, 2002, indicate that the federal government will possibly expand the number of individuals vaccinated to 500,000 first responders and emergency room personnel.[4] However, other media sources indicate that reports of such an expansion in the numbers of those vaccinated are inaccurate; HHS will release its implementation plan by the end of July.[5] States could begin receiving vaccine as early as September.

Increasing the number of healthcare personnel vaccinated may be in response to criticism of the ACIP proposal. The Executive Director of the American Public Health Association, Mohammad Akhter, stated that in the event of a smallpox outbreak, victims would head to the nearest hospital rather than a ACIP designated hospital. He was also concerned that the small number of designated hospitals would become overwhelmed by such patients, and therefore advocated that all emergency room personnel should be vaccinated. The Federation of American Hospitals expressed similar concerns, and supports vaccinating more health care workers than under the ACIP plan.[6]

Federal officials are hesitant to recommend vaccination of the general public because smallpox vaccine has a higher complication rate than any other vaccine in current use. An estimated 28 million people with eczema and 10 million people with a compromised immune system (organ transplant recipients, HIV/AIDS or cancer patients), or 15% of the population, are at high risk for developing serious complications.[7] Such complications will also occur in the rest of the population, but at a lower rate. Liability problems associated with smallpox vaccine could be enormous. The serious complications include:

- postvaccinial encephalitis may be fatal or leave survivors with paralysis or other central nervous system symptoms;
- progressive vaccinia, or growth of the vaccination lesion without healing, occurs in immunocompromised individuals and is often fatal;
- fetal vaccinia, passage of vaccinia to a fetus, may lead to stillbirth;
- eczema vaccinatum, or development of vaccinial lesions over sites where there is or has been eczema, may be fatal.

Complications may occur in individuals who don't even receive the vaccine, but are only exposed to someone who has recently been vaccinated. In fact, eczema vaccinatum is more severe among unvaccinated persons exposed to a vaccine recipient than among recent vaccine recipients. One recent study found that even if high risk individuals and their household contacts (25% of the U.S. population) were excluded from vaccination, an estimated 1600 serious adverse events and 190 deaths would occur if people aged 1 to 29 years were vaccinated, and 4600 serious adverse events and 285 deaths would occur if people aged 1 to 65 years were vaccinated.[8] Less serious complications of vaccination include rash, fever, and

---

3 The draft recommendations can be found at: [www.cdc.gov/nip/smallpox/supp_recs.htm].

4 Broad, W. U.S. to Vaccinate 500,000 Workers Against Smallpox. *The New York Times*, July 7,2002. p. A1, A16.

5 Hawana, J.S. Advance Smallpox Vaccination Recommended in PNAS Report; HHS plan due by end of July. *Washington Fax*, July 9, 2002.

6 Young, J. Proposal to Designate Smallpox Response Center Hospitals Evokes Concerns. *Washington Fax*, June 26, 2002.

7 Kemper, A.M., et al. Expected Adverse Events in a Mass Smallpox Vaccination Campaign. *Effective Clinical Practice*, v. 5, March/April 2002. p. 84-90.

8 Ibid.

accidental inoculation, in which the rash occurs elsewhere on the body (from the vaccination site) due to direct contact with vaccinia. Blindness may result if the rash develops near the eye.

Federal vaccination policy must balance the certainty of the adverse effects of smallpox vaccine against the uncertainty of a biowarfare attack using variola virus. In the past, naturally occurring outbreaks of smallpox have been controlled using ring vaccination: all the patient's contacts are tracked down, vaccinated and quarantined until the disease spread is stopped. Such a strategy is logical if there is limited vaccine, or a significant level of smallpox immunity in the population, or if a society wishes to limit the number of adverse events and fatalities caused by mass smallpox vaccination. However, in a modern society with low levels of smallpox immunity, mass transit and rapid forms of long distance travel, ring vaccination may not be feasible especially if the attack occurs in many sites and is relatively large. Moreover, our social and economic system may be greatly stressed by the necessary implementation of quarantine.

Researchers are using mathematical models of disease transmission to determine the outcome of a smallpox attack under a range of different conditions, such as the number of initial infections, the transmission rate (number of secondary cases caused by each initial case), the number of vaccinators, and whether ring or mass vaccination is used to control the outbreak. The models are still under development and provide contradictory results. For example, a model developed by researchers at Emory University found that "ring vaccination, even when it's started only after the 25th case of smallpox, can contain an epidemic almost as well as mass vaccination, provided that at least 80% of those exposed can be found and vaccinated."[9] In contrast, a model developed at Yale University found that "mass vaccination results in far fewer deaths and much faster epidemic eradication."[10] The Yale model predicts that in a smallpox attack on a large city causing 1000 initial cases, ring vaccination results in 367,000 cases and 110,000 deaths over 350 days, but mass vaccination following an attack results in 1,830 cases and 560 deaths over 115 days. However, if 40% of the population is vaccinated prior to attack, the number of deaths under ring vaccination and mass vaccination are 40,000 and 440 respectively. The Yale group concludes "that unless pre-attack vaccination is used ..., serious consideration should be given to replacing the existing CDC policy [with] mass vaccination in the event of a smallpox attack in a large urban center."

Traditional methods, like ring vaccination, are based on the fact that smallpox spreads through person-to-person contact. A controversial report, presented at a June 2002 meeting on smallpox vaccination policy, indicates that aerosolized smallpox virus from a Soviet bioweapon test may have accidently caused 10 cases of smallpox and three deaths in 1971 in what is now Aralsk, Kazakhstan.[11] Previously, scientists believed that aerosolized smallpox virus would be inactivated by environmental factors, such as UV radiation from the sun. However, the authors believe the best explanation for the outbreak is that the initial case acquired smallpox when her boat entered waters downwind from a Soviet bioweapons test site located on an Aral sea island. Because an aerosol of smallpox virus would have the potential to infect a larger number of people than person to person contact, this chapter is of concern to public health experts.

---

[9] Enserink, M. How Devastating Would a Smallpox Attack Really Be? *Science*, May 31, 2002. p. 1592-1595.

[10] Kaplan, E.H. et al. Emergency Response to a Smallpox Attack: the Case for Mass Vaccination. *Proceedings of the National Academy of Sciences, Early Edition*, July 8, 2002. [www.pnas.org]

[11] Enserink, M. Did Bioweapons Test Cause a Deadly Smallpox Outbreak? *Science*, June 21, 2002. p. 2116-2117.

Although the 119 million unvaccinated people in the U.S. would face a mortality rate of about 30% in a smallpox attack, there is some question over the level of protection remaining in the 157 million individuals who were vaccinated more than 30 years ago. Complete protection from smallpox disease begins to taper off 3 to 5 years after vaccination. However, some historical data indicate that vaccinated individuals may experience milder disease symptoms and a lower mortality rate (5%) even up to 50 years post-vaccination.[12] Additional studies would be needed to confirm this point, but if correct, a smallpox attack might not be as deadly or spread as rapidly as some experts had thought. It might also influence decisions on who in the U.S. population should be the first to receive vaccines if a smallpox attack does take place.

Within the last year, HHS has been successful in greatly increasing the U.S. stockpile of smallpox vaccine via agreements with three separate sources: Wyeth Laboratories, a pharmaceutical company located in Marietta, PA; Acambis, a British drug firm with offices in Cambridge, MA; and Aventis Pasteur, a French vaccine company with a plant in Swiftwater, PA. The policy debate by ACIP and others on how many U.S. residents should be vaccinated has been made possible by HHS efforts to increase the U.S. stockpile of smallpox vaccine.

In the fall of 2001, the National Institute of Health (NIH) funded a study on whether the 15.4 million doses of smallpox vaccine in the U.S. stockpile could be diluted without losing its potency. The vaccine, called Dryvax, was manufactured by Wyeth in 1982. Results of the study, released in April 2002, indicate that the Dryvax can be safely diluted by a factor of 5 or 10. In 97% to 99% of study participants the diluted vaccine produced a "take" or skin lesion, a positive indicator of a protective immune response.[13] Although none of the 680 study subjects, aged 18 to 32, had a serious adverse reaction, there was a high frequency of pain, swelling and redness at the inoculation site as well as fever headache, rash, muscle aches, fatigue and chills. More than a third of the subjects missed work, school, sleep, or other activities because of an adverse reaction.

In November 2001, HHS placed a rush order for an additional 155 million doses of smallpox vaccine (beyond an initial order of 54 million doses) with Acambis. Clinical trials of the vaccine began in March 2002.[14] The vaccine is a single strain of vaccinia, in contrast to the mixture in Dryvax, and is produced in a cell culture rather than the traditional method which used cows or calves. Although animal studies indicate that the Acambis version may have a lower risk of causing encephalitis, the other side effects are expected to be about the same as the traditional vaccine. Acambis expects to deliver the vaccine order by the end of 2002. At the end of March 2002, Aventis announced that it would donate to the U.S. government about 85 million doses of smallpox vaccine that were manufactured in 1958 for the U.S. Department of Defense. Preliminary tests of the vaccine indicate that it is probably as potent as Dryvax.[15] Clinical trials of the Aventis vaccine began in the spring of 2002. Dilution tests of the vaccine will also be conducted.

---

[12] Cohen, J. Smallpox Vaccinations: How Much Protection Remains? *Science*, v. 294. November 2, 2001. p. 985.

[13] Frey, S.E. et al. Clinical Responses to Undiluted and Diluted Smallpox Vaccine. *The New England Journal of Medicine*, v. 346, April 25, 2002. p. 1265-1273.

[14] Enserink, M. Smallpox Vaccines: New Cache Eases Shortage Worries. *Science*, v. 296, April 5, 2002. p. 25, 27.

[15] Ibid., p. 25.

CDC is negotiating a contract to increase the U.S. stockpile of a product used to treat serious adverse reactions to smallpox vaccine.[16] The product, vaccinia immune globulin intravenous treatment (VIGIV) is expected to be available in 2003. Currently CDC has access to a similar product that could be used to treat adverse reactions in about 675 adults, the number expected when vaccinating 4 to 6 million people. CDC is seeking to increase the stockpile by 30,000 doses, the amount needed if a mass U.S. smallpox vaccination did not include pregnant women. If pregnant women were included, an additional 40,600 doses would be needed.

If the smallpox vaccine could be used without the high rate of serious complications, the threat of a bioterrorism weapon using smallpox virus would be eliminated. The federal government is encouraging both academic scientists and companies to develop a safer vaccine. Aventis Pasteur plans to begin testing a smallpox vaccine using a greatly weakened strain of vaccinia called NYVAC.[17] A combination of the traditional vaccine Dryvax with a primer vaccine called modified vaccinia Ankara (MVA) is also being studied. MVA is given several months before Dryvax and was used in German studies in the 1970s and recent studies by Bavarian Nordic of Copenhagen to eliminate serious side effects in healthy and high risk individuals. However, because MVA was never tested where smallpox virus was endemic and MVA with Dryvax does not produce a typical smallpox vaccine scar, there are concerns that it may not provide adequate disease protection. A monkey model of smallpox, developed by CDC and the U.S. Army Medical Research Institute of Infectious Diseases (USAMRIID), will be used to test the efficacy of MVA. The Food and Drug Administration (FDA) amended its new drug and biological product regulations so that products intended to prevent serious or fatal conditions may be approved for use based on effectiveness data from animal studies when human efficacy studies are not ethical or feasible. The new regulations took effect on June 30, 2002.[18]

Scientists at CDC and USAMRIID are working on antiviral treatments for smallpox. One drug, cidofovir, has demonstrated activity against variola virus in preliminary tests. In 2001, an Investigational New Drug (IND) application was filed with FDA for the "use of cidofovir in both the treatment of acute smallpox infection and the management of adverse events associated with vaccinia immunization."[19] Other compounds 25 to 150 times more active than cidofovir have been identified and are under investigation.

LEGISLATION

The Public Health Security and Bioterrorism Preparedness and Response Act of 2002 (P.L. 107-188), signed by the President on June 12, 2002, directs the Secretary of HHS to ensure that there is enough smallpox vaccine in the Strategic National Stockpile to meet health security needs and authorizes $509 million for FY2002 and such sums as may be necessary through FY2006 for this purpose. H.R. 5005, the Homeland Security Act of 2002

---

16 Smallpox Vaccine Adverse Event Treatment Will Be Made Under CDC Contract. *The Pink Sheet*, July 1, 2002. p. 17-18.

17 Enserink, M. In Search of a Kinder, Gentler Vaccine. *Science*, v. 296, May 31, 2002. p. 1594.

18 *FDA News*, May 30, 2002, available at: [www.fda.gov/bbs/topics/news/2002/new00811.html]

19 LeDuc, J.W. et al. Smallpox Research Activities: U.S. Interagency Collaboration, 2001. *Emerging Infectious Diseases*, v. 8, July 2002. p. 743-745.

as passed by the House, would amend P.L. 107-188 by moving authority for the stockpile to the Department of Homeland Security (DHS); HHS would continue to manage the stockpile and determine its contents. S. 2452, the National Homeland Security and Combating Terrorism Act of 2002, as reported by the Governmental Affairs Committee on July 25, 2002, directs DHS to consult with CDC in the administration of the stockpile by CDC. An emergency supplemental appropriation (P.L. 107-117), signed by the President on January 10, 2002, provided $512 million for the purchase of smallpox vaccine by HHS.

In: Focus on Bioterrorism
Editor: George T. Foster, pp. 25-37
ISBN 1-60021-185-2

Chapter 3

The U.S. Postal Service Response to the Threat of Bioterrorism through the Mail

Frank Gottron

Abstract

The deliberate mailing of *Bacillus anthracis* spores through the U.S. Postal Service (USPS) has caused five deaths, twenty-two cases of anthrax, and massive disruptions to Congress and the USPS.

Both the public and private sector are examining an array of methods to limit the risk of future attacks. The array of potential solutions include improving mail handling procedures, changing the USPS anonymous mailing policy, installing bio/chem agent detectors, and sterilizing the mail.

For the USPS these decisions are complicated by its precarious financial state. Some proposed solutions may require an increase in postage rates and/or decreased levels of service. Each of these may further depress postal revenues and threaten the continued existence of the USPS as an independent, self-supporting entity.

Policymakers may wish to consider if the USPS must ensure the safety of mail recipients. At this point it is not clear if this is practical or even possible with existing technology. It may be that it is practical to protect only the mail addressed to the most likely targets of future attacks.

Some of the measures that the USPS has taken or is planning to take to protect postal workers and mail recipients are common sense alterations to the mail processing procedures. These include measures to reduce cross contamination such as using vacuuming instead of pressurized air to dislodge dust and adding filters to air handling systems. The USPS is studying the feasibility of including biological weapon detectors during the sorting process. Additionally, the USPS has made gloves, masks, and educational materials available to all postal workers.

More controversial and potentially more costly are plans to sterilize the mail. Currently, all mail destined for federal offices in the Washington DC metropolitan area is shipped to a sterilization facility for irradiation treatment before delivery. The USPS is studying whether this solution can be scaled up to sterilize all mail from anonymous senders. To implement irradiation procedures nationwide could cost up to $2.25 billion with up to another billion dollars each year in operating costs. This procedure will cause delays and may damage the contents of some mail.

Also controversial and costly are plans to increase product tracking systems that will include video images of all retail transactions. The planned system will enable mail dropped off at post offices to be tracked through to delivery and linked to facial images of senders. Mail dropped off at anonymous collection sites will be traceable to the drop off point. Stamps and envelopes could also be encoded with purchase location information.

Policymakers face a difficult balance to strike between concerns for safety, cost, and practicality while deciding how to alter the practices of the USPS.

INTRODUCTION

The use of the U.S. Postal Service (USPS) to deliver agents of bioterror has caused 22 confirmed cases of anthrax including five deaths as well massive disruptions to Congress and the USPS.[1] The contamination caused by processing and opening of the letters shuttered the Hart Senate Office Building for more than three months and indefinitely closed mail processing centers in Washington DC and Trenton, NJ.

These attacks have radically changed the way the nation views mail delivery and mail safety. Both the public and private sectors are evaluating a variety of methods to reduce the risk of similar future attacks. These potential solutions include changing mail handling procedures, changing the USPS anonymous mailing policy, installing bio/chem agent detectors and sterilizing the mail.

For the USPS these decisions are complicated by its serious financial difficulties.[2] The USPS projects costs of $762 million in FY2002 for clean up costs, medical treatment for postal workers, measures to decrease the threat of future mailings, protect postal workers, recovery costs medical treatments, mail irradiation, and to begin evaluating systems to protect mail recipients. In the face of these increased costs, the USPS predicts the decrease in mail volume caused by the attacks will cost $2 billion dollars in lost revenues this fiscal year. USPS feels that its survival depends on being able to ensure the health of postal workers and to restore the faith of the American people in the safety of the mail.

INTEREST AND ROLE OF CONGRESS

The efforts of the USPS to respond to the threat of bioterrorism has been the subject of hearings before the Senate Treasury and General Government Appropriations Subcommittee, Senate Committee on Governmental Affairs and the House Committee on Government Reform. At the request of Committee on Government Reform Chairman Dan Burton and Ranking Member Henry Waxman, the National Academy of Sciences and the General

1 Centers for Disease Control and Prevention. Morbidity and Mortality Weekly Report. *Update: Investigation of Bioterrorism-Related Anthrax---Connecticut, 2001.* December 7, 2001. Vol 48. No. 48. p. 1077. [http://www.cdc.gov/mmwr/preview/mmwrhtml/mm5048al.htm].

2 For a thorough review of the USPS financial situation, see CRS Report RL31069, *Postal Service Financial Problems and Stakeholder Proposals*, by Nye Stevens. [http://www.congress.gov/erp/rl/pdf/RL31069.pdf].

Accounting Office have convened meetings between USPS officials and outside experts to discuss appropriate technology and methodological changes to limit the threat of future attacks.

Postmaster General John Potter has stated that he views the need to reduce the threat of terrorism through the mail as an integral part of homeland defense and thus plans to request Congress to appropriate billions of dollars.[3] In addition to the $175 million it received from the White House Emergency Response Fund, the USPS initially estimated that it would need an additional approximately $1.1 billion in FY2002 to respond effectively to the bioterror threat and to replace or repair facilities damaged in the September 11, 2001 terrorist attacks. The 2002 Department of Defense Appropriations Act (PL 107-117) appropriated $500 million to the USPS for these expenses. Because of ongoing concerns over plans to screen and sanitize mail, the Conferees required the USPS to submit an emergency preparedness plan to the Committees on Appropriations, the House Committee on Government Reform and the Senate Committee on Governmental Affairs before the money for these activities would be disbursed.

> ...As part of its emergency preparedness plan, the conferees expect the Postal Service to include an assessment of threats to the health and safety of employees and customers of the Postal Service and the integrity of the mail; testing and evaluating the options for detecting and/or addressing those threats, including both technology-based and process-based options; a comparison of the costs and benefits of options under consideration; an evaluation of the strengths and weaknesses of the technologies under consideration for mail sanitization, including an analysis of risks to human health and safety and to mail products associated with each of those technologies; and a timetable for implementing the options selected....[4]

In response to this request, the USPS published its comprehensive emergency plan in March 2002[5]. This document discusses how the USPS has responded to the attacks and the programs it plans to pursue over the next five years to further reduce the chances of a successful bioterror attack.

In addition to the $675 million it has already received from Congress, the USPS indicated that it may ask for an additional $4.1 billion over the next five years to implement this plan. Because of its high cost, this plan is likely to fall under careful scrutiny. Policymakers may want to explore questions that define the role the USPS should play. Clearly the USPS must protect postal workers, but should the USPS also take responsibility for the safety of mail recipients? Is that possible? If it is possible, how much are ratepayers or taxpayers willing to pay? Does safe mean safe from all potential biological attacks or only anthrax? What about chemical agents? What kind of reductions in USPS service are acceptable to ensure safety? Will changes to level of service add to the precarious financial position of the USPS? What

---

3 Testimony of Postmaster John Potter before the Senate Treasury and General Government Appropriations Subcommittee November 8, 2001.

4 U.S. Congress. House. *Making Appropriations for the Department of Defense for the fiscal year ending September 30, 2002 and for other purposes.* 107th Congress, 1st session. H. Rept. 107-350, Washington, U.S. Govt. Print Off. 2001. p. 452.

5 USPS. *The U.S. Postal Service Emergency Preparedness Plan for Protecting Postal Employees and Postal Customers from Exposure to Biohazardous Material and for Ensuring Mail Security Against Bioterror Attacks.* March 6, 2002. [http://www.usps.gov/news/2002/epp/welcome.htm].

portion of the costs should Congress appropriate and how much of the costs should rate payers bear?

Initial Response

Previous terrorist attacks via the mail, such as those of the Unabomber, generally targeted the recipient of the mail. However, because anthrax spores are smaller than the natural pores in envelopes, the spores can leak out during mail sorting, putting postal workers at risk. Additionally, spores can be transferred to other envelopes (a phenomenon called cross contamination), which puts at risk anyone whose mail was processed at the same facility as the anthrax tainted envelopes. Before these attacks, these risks were generally considered negligible by experts in biological warfare and the Centers for Disease Control and Prevention (CDC) because it was thought that the few spores that could be transferred this way would not be sufficient to cause disease. Because this assumption proved wrong, the initial response of the USPS has focused on ways to protect postal workers from exposure and how to protect a subset of high risk addressees.

Shortly after the discovery of the contaminated letters, both mail processing facilities which handled the letters were closed indefinitely. The approximately 8,500 postal workers deemed at even remote risk for exposure to anthrax were offered a 60-day course of prophylactic antibiotics.[6] Following the initial 60-day course of antibiotics the CDC has recommended an additional 60-day course. The CDC also has made the controversial anthrax vaccine available for the workers.

The delay between the discovery of the tainted letters, facility closures and antibiotic treatments elicited criticism both from postal worker unions and Congress.[7] The USPS attributed the difference between the immediate closing and testing of workers in the Hart Senate Office Building and the delay before postal facilities and workers were tested to a reliance on the evolving recommendations of the CDC.

After these attacks, the USPS examined its practices and found some changes could be made to reduce the risk to both postal workers and mail recipients. One of the most obvious changes concerned dust abatement procedures. Processing mail generates an enormous amount of dust. Compressed air had been the method of choice to clean the processing machines. The USPS recognized that this practice would widely distribute fine anthrax particles throughout the processing facility. It is switching to vacuum systems both to clean the sorting machines and to replace the general maintenance practice of broom sweeping. These vacuums are fitted with high efficiency particulate air (HEPA) filters that are designed to capture very small particles such as the bacterial spores that cause anthrax. The USPS has also provided antimicrobial wipes, protective gloves and masks, and their related educational materials to ensure proper use to all postal workers.

Prior to the discovery of the tainted mail, no procedure was in place to sterilize potentially harmful biological agents in the mail. As an interim measure the USPS has decided to sanitize mail destined for all federal government offices in the Washington D.C.

---

6 Centers for Disease Control and Prevention. *Morbidity and Mortality Weekly Report.* Vol.50 No. 47, November 30, 2001, p. 1052.

7 Washington Post.*U.S. Officials Defend D.C. Response.* October 24, 2001. p. A16.

metropolitan area (zip codes 20200-20599). Initially this consisted of sealing the mail into secure containers and shipping it to subcontractors in Lima, Ohio and Bridgeport, New Jersey to be sterilized by irradiation. The USPS claims to have sterilized and delivered all of the mail that had been held undelivered since October, although some agencies including the Library of Congress contend that all of this mail has not yet been delivered.[8] Once the undelivered mail backlog was cleared, the USPS began shipping all the mail only to the Bridgeport facility. This sterilization procedure adds approximately five to seven days to delivery times. The USPS forecasts that this procedure will be in place for only six months while a permanent solution is implemented which could reduce the delay to two days.

The USPS has purchased eight sterilization machines from the Titan Corporation for a total of $40 million. These will be used as part of an interim solution to sterilize the mail of what is considered high probability targets. These targets were determined by the USPS in cooperation with the Department of Justice. For national security reasons, the USPS will not disclose the exact make up of this list. However, they have publicly stated that some of these initial eight machines will be used to sterilize the mail destined for government offices in Washington D.C. If the decision to sterilize all mail is made, it may take several years before all of the required equipment can be installed. In this case, the target list will be used to prioritize which facilities to upgrade first.

The cost of some of this initial response was covered by the $175 million received from the White House Emergency Response Fund (Table 1.) Other aspects of the initial response, such as the HEPA filtration systems, are to be covered by the $500 million included in the 2002 Department of Defense Appropriations Act supplemental appropriation (Table 2.) In addition, the $87 million that the USPS indicated was still needed to cover prevention and protection initiatives in FY2003 (Table 3) is part of the addition supplemental FY2002 appropriations. This passed the House (H.R. 4775) on May 24, 2002 and is under consideration of the Senate (S. 2551.)

LONG TERM PLAN

The future viability of the USPS may depend on its long term response to this new challenge. The USPS has decided while it is impossible to completely eliminate all risk of a successful bioterror attack through the mail, a multifaceted response can greatly increase the safety of postal workers and mail recipients. The USPS constructed the Emergency Preparedness Plan around four objectives: deterrence of using the mail as a tool for bioterrorism, and the detection, containment and neutralization of biohazardous materials in the mail stream as soon as possible. To accomplish these goals the USPS proposes six initiatives: Prevention; Protection and Health-Risk Reduction; Detection and Identification; Intervention and Sanitization; Decontamination; and Investigation. Because several aspects of this plan are still under study, the USPS describes it as "dynamic" and subject to change. Only the Decontamination initiative has been completely funded by the appropriations to

---

[8] Letter from James Billington, Librarian of Congress to Postmaster General John Potter, April 3, 2002. See also, Gail Fineberg. *Library Seeks Mail Missing Since October.* The Gazette. Vol. 13. No. 14. Library of Congress. April 12, 2002. [http://www.loc.gov/staff/gazette/pastissues/2002/041202/htdoc/gzst002.html]

date. The estimated costs of the remaining five as yet unfunded initiatives are detailed in Table 3.

Table 1. Usage of the White House Emergency Response Funds

	Cost as of 2/28/2002 ($ in millions)
Irradiation Equipment and Services	53.0
Employee Personal Protection	29.0
On-Site First Response Environmental Testing	24.0
Nationwide Mailing and Communications	15.0
Site Clean Up	45.0
Medical Costs	9.0
Total	**175.0**

Source: USPS. The U.S. Postal Service Emergency Preparedness Plan for Protecting Postal Employees and Postal Customers from Exposure to Biohazardous Material and for Ensuring Mail Security Against Bioterror Attacks. March 6, 2002.

Table 2. Usage of the $500 Million in FY 2002 Supplemental Funds

	Cost ($ in Millions)
On-going Activities	
Decontamination	35.0
Proposed Activities	
Threat Detection and Identification	200.0
Sorting Machines and Maintenance Vacuums HEPA Filter Retrofit	254.0
Emergency Response Plan Preparation	0.5
Facility Repai	10.5
Total	**500.0**

Source: USPS. The U.S. Postal Service Emergency Preparedness Plan for Protecting Postal Employees and Postal Customers from Exposure to Biohazardous Material and for Ensuring Mail Security Against Bioterror Attacks. March 6, 2002.

Prevention

The USPS plan defines prevention to include programs that stop contaminated mail from entering the mail stream as well as efforts to reduce the attractiveness of using the mail as a delivery system for bioterror agents. The Emergency Preparedness Plan focuses on redesigning collection boxes, increasing security at retail outlets, developing security procedures for commercial mailers, and increasing security at truck entrances.

Table 3. Supplemental Funding Requirements of USPS Emergency Plan

	Costs ($ in Millions)			
	FY2002	**FY2003**	**FY2004**	**FY2005-2006**
Prevention				
Collection Box and Drop Slot Modification*	10.0	50.0	292.0	
Retail Outlet Modification*		150.0	100.0	
Increasing Security at Truck Entrances		135.0		
Protection and Health-Risk Reduction				
Sorting Machines	71.0	164.8	30.0	
HVAC*	6.0	120.0		
Threat Detection and Identification				
Threat Detectors and Identifiers		180.0	120.0	
Intervention and Sanitization				
Mail Irradiation Facilities				1,000 – 2,250
Investigation				
Image Capture			24.0	
Image Analysis			112.0	
Mail-Piece Tracking			210.0	
Positive Product Tracking			9.5	111.0
Total	**87.0**	**799.8**	**897.5**	**1,111 – 2,360**

* Estimated costs. Actual funding needed depends on outcome of further USPS study.

Source: USPS. The U.S. Postal Service Emergency Preparedness Plan for Protecting Postal Employees and Postal Customers from Exposure to Biohazardous Material and for Ensuring Mail Security Against Bioterror Attacks. March 6, 2002.

Collection Box and Drop Slot Redesign

Immediately following the anthrax outbreak, some people called for the removal of all collection boxes. The USPS considered this, but decided that such a step would produce an enormous drop in residential and small business patronage. The USPS concluded that removal of all collection boxes was inconsistent with its continued operation. Therefore attention was focused on minimizing the risk associated with anonymous mail drops.

Currently, when mail enters one of the 350,000 collection boxes or 40,000 post office drop slots, it falls into a three-sided bin. The USPS proposes to redesign these to have the mail enter a plastic bag which would be mechanically sealed before the postal worker collected the mail. The USPS is studying the feasibility of including a disposable biohazard detector in each bag to immediately identify dangerous mail. This combination would prevent exposing postal workers to bioterror agents and detect contaminated mail before it enters the mail stream. Although currently available biodetection strips are unlikely to be sensitive enough for this use, the USPS believes that either the next generation of strips will be useful or that another inexpensive detector system may become available soon.

Collection boxes and mail bags may also be modified to contain tracking information. Thus, if a tainted letter gets past the initial detector screen but is detected later, the entire route of the letter could be traced, which would help health care, clean up, and investigative operations.

Increasing Retail Outlet Security

This initiative calls for extensive deployment of security cameras to record all retail transactions. Images would be stored locally and could be linked with mail piece tracking to increase the investigative abilities as well.

Critics of this initiative may suggest that $250 million is too much money for a system that a terrorist can easily thwart by avoiding post offices. Critics may also be concerned about the privacy issues raised by videotaping all postal patrons. The USPS acknowledges that this initiative may change public perceptions of its role, which may further depress revenue.

Developing Security Procedures for Commercial Mailers

Because two-thirds of mail comes from commercial bulk mailers, the USPS will work with industry to develop security standards to reduce the probability of receiving contaminated mail from this source. This mail will be considered "safe" and not subjected to further detection or possible decontamination processes. Currently there is no request for supplemental funding associated with this initiative because such a program has not yet been developed. However, the USPS predicts that any costs will be relatively low, with industry paying for the bulk of it while the USPS will hire compliance personnel.

Increasing Security at Truck Entrances

This initiative will upgrade and install mechanical and electronic security devices and add more security personnel at each of the approximately 400 large mail processing centers.

Protection and Health-Risk Reduction

Sorting Machines

It is likely that mail became cross contaminated inside the processing machines when the pinch belts, used to propel mail, squeezed air out of the envelopes and caused a puff of spores. The USPS is in the process of designing HEPA filtration systems for these sorting machines. The installation of these filters should be completed by the end of FY 2004 and is expected to require an additional $265.8 million in supplemental appropriations (Table 3.)

Heating, Ventilating, and Air-conditioning Systems

The USPS is also studying the use of HEPA filters in the heating, ventilating, and air-conditioning systems (HVAC) all of the mail processing centers. The USPS projects that these systems could be operational by the end of FY2003 for a total cost of $126 million. Because the HVAC for each processing center is unique, a large portion of the implementation cost arises from the need for to design an individual system for each site. The USPS is also considering using an air sterilization system either instead of or in addition to HEPA filter retrofits. It may be possible to use ultraviolet light or ultrasound to kill any spores passing through the ventilation ducts. However, critics of HVAC modifications argue that only a small percentage of the air in a facility actually passes through these systems and therefore spores could stay airborne for days or weeks before entering a duct for filtration or sterilization.

Detection and Identification

The USPS envisions a system which can continuously monitor facility air and the air ejected from mail during processing for potential threats. Current plans call for a two step system, detection of a likely threat and confirmation/identification of the agent. Following confirmation, appropriate steps could be taken such as shutting down the center and treating at-risk personnel.

Particle Analysis

The first step, also called the triggering step, would likely be carried out by a particle analyzer. These machines continuously sample the air and measure the size of all airborne particles. More sophisticated versions also measure the aerodynamic shape of the particles and can potentially discriminate between biological and nonbiological particles. These machines send an alert when it detects a change in particle number, size, shape or makeup from background levels. Adapting this technology to USPS needs is complicated due to the very high levels of paper dust that is generated during mail sorting. However, because paper dust particles tend to be larger than spores, it is likely that this technological hurdle can be overcome with appropriate filtering. Because this type of detection system has a relatively high false alarm rate, a second step is required for verification.

An alarm by the particle analyzer would trigger a second step to verify the presence of a threat and identify the agent. It is not clear what technology would be used for this step. Experts from the Department of Defense advised the USPS that the following technologies might be best suited for this application: biological indicator strips, polymerase chain reaction (PCR), and mass spectroscopy.

Biological Indicator Strips

Biological indicator strips work like a home pregnancy tests. A sample is placed on the strip and within minutes a result is known. This can be automated to fit with USPS plans. However, many experts think that biological test strips might require the presence of high levels of agent to work reliably. Therefore potentially lethal amounts of agent might not be detected. Another drawback is that each agent requires its own specially prepared test strip. Therefore samples can only be tested for only a finite number of known agents.

Polymerase Chain Reaction

PCR is a technology for detecting small quantities of DNA with a particular genetic sequence. Typically, tests are designed to be specific for a given species or strain of an organism. If the target DNA sequence is present in any of the DNA in the sample, the reaction produces multiple copies of the target sequence. These multiple copies can then be detected to give a positive test result. If no target sequence is present, no copies are made, and a negative test result is obtained. Multiple agents can be tested for simultaneously using PCR. PCR can detect much lower levels than biological test strips and it is relatively easy to expand the list of agents for which to test. However this test only works for agents which have genetic material. For example, PCR will detect bacteria such as the one that causes anthrax but it will not detect chemical agents nor biological toxins such as botulinum toxin.

Mass Spectroscopy

Mass spectroscopy is an advanced technology that is thought to be the "gold standard" for chemical detection. It breaks down samples into their constitutive parts then determines the mass of each piece. The instrument produces a complicated graph, or spectra, that shows the masses of all of its parts. This spectra is unique for each agent. Chemical agents have relatively simple spectra and are easily identified by this method. In theory, this method will work equally well on biological agents. However, because biological agents have many more constituent parts, they produce much more complicated patterns. This makes discriminating between a dangerous bacteria and a harmless related organism very difficult. Although this technology may currently exist, it is unlikely be in a system that could be adapted quickly for the USPS to use. As this technology develops, it may become the method of choice for these types of applications, but some experts feel that this is at least a few years away.

Intervention and Sanitization

Intervention procedures are used to neutralize potential mail contaminants for mail destined for high risk targets. As described above, this group currently consists of all federal government offices in the Washington D.C. metropolitan area (zip codes 20200-20599). After evaluating various forms of mail sanitization including gaseous and radiation methods, the USPS chose electron beam irradiation.[9] Electron beam, or e-beam, is essentially a very powerful verison of the electron gun found in the cathode ray tubes of televisions and computer monitors. It is a commonly used method to sterilize medical supplies including baby bottle nipples and bandages.

The USPS is leaving open the option of expanding mail irradiation to include all noncommercial mail. However, in light of the problems generated by the current limited use[10], this appears increasingly unlikely. Recipients have complained of various adverse health reactions to the irradiated mail as well as damage to mail contents. The USPS claims that lowering the radiation dose has solved many of these problems. The Emergency Plan predicts that a system-wide irradiation program would require $1-2.25 billion in capital costs (Table 3.) Some experts suggest that annual operating costs for this system could run as much as $1 billion. If implemented the USPS would likely seek an appropriation for the capital costs while relying on rate payers for the operating costs. This plan could be implemented as early as FY2005.

---

[9] For more information about mail irradiation, see CRS Report RS21184 *Irradiated Mail* by Frank Gottron. [http://www.congress.gov/erp/rs/pdf/RS21184.pdf]

[10] Ibid.

Decontamination

These procedures differ from intervention procedures in that they would be used following the identification of a contamination event rather than prophylactically. This covers treatment of contaminated mail and facilities. Tainted mail was treated using e-beam irradiation. The USPS evaluated several methods to treat tainted equipment and facilities before deciding on chlorine dioxide. This is the same technology used on the Hart Senate Office Building. The $80 million required for decontamination came from both the $175 million White House Emergency Response funds (Table 1) and from the $500 million supplemental appropriation (Table 2.)

Investigation

These procedures are designed to help the USPS investigate successful mailbased terrorist attacks. The technologies under consideration are based on letter image capture and analysis. These improvements are predicted to cost $355.5 million in FY2004 and $111.0 in FY2005.

Image Capture

The USPS is currently in the process of upgrading their ability to digitally capture an image of every mailpiece as it is sorted. Current capabilities only allow a portion of the address side to be captured. Deployment of the Wide Field of View Image Camera will allow the entire side to be imaged. The USPS considers this to be crucial for the ability to reliably analyze the images.

Currently, the images of each envelope are deleted immediately after being used for sorting purposes. The Emergency Plan calls for $24 million in FY2004 to increase storage capabilities and complete deployment of the new cameras so that images can be stored for 30 days.

Image Analysis

The USPS also plans to implement programs that could profile the images of mail-pieces in real-time. Some of this analysis could redirect mail that met specific suspicious criteria for special handling. It may also be possible to analyze handwritten addresses to search for and isolate mail-pieces currently in the system that satisfy specific criteria. For example, letters that share distinctive handwriting such as that found on the anthrax related mailings might be able to be matched automatically and redirected for special handling. The USPS predicts these projects will cost $112 million and be available in FY2004.

Mail-Piece Tracking

The Emergency Plan calls for an expansion of the USPS' current mail tracking capabilities by linking a unique identification barcodes to each envelope image. This would allow investigators to trace the exact path through the system of any contaminated mail. This is expected to cost $210 million and be implemented in FY2004.

Positive Product Tracking

This initiative would result in an increased ability to track mail. This system would encode all products with a specific identifier that would be traceable to specific retail transaction and video image of the patron. This includes any mail dropped off at a retail outlet and any products bought, such as stamps or envelopes. Furthermore, all USPS products could be encoded to allow investigators to determine the location of its sale (for example, the kiosk at which a specific stamp was purchased.) This information could be useful even without a linked video image of the patron that would be obtained if the transaction took place in a post office. The USPS sees this as complementing and extending the abilities discussed under retail outlet modifications. The USPS predicts that these modifications would cost $9.5 million in FY2004 and $110 million in FY2005.

Critics say this initiative raises privacy questions and could result in serious public relations difficulties for the USPS. They argue that this will cost much more that projected capital costs because of the revenue lost due to decreased patronage by concerned customers.

Conclusions

The Emergency Preparedness Plan outlines a series of steps that the USPS has taken and is planning to take to decrease the threat of biological terrorism through the mail. Some of the steps, such as the change to HEPA filtered vacuums, are noncontroversial and easily implemented. Other steps, such as threat detectors are expensive and fraught with technical challenges. However the Plan also includes some expensive and very controversial steps such as the videotaping and tracking of all retail transactions and the irradiation of all noncommercial mail.

The USPS stresses that this plan is dynamic and will most likely change as new technologies become available. However, one thing that policymakers will have to decide is how much of these plans should be paid by taxpayers rather than ratepayers. If fully implemented, the USPS would seek up to $4.1 billion in additional appropriations for FY2003 through FY2006. Critics who feel that the USPS was financially doomed before the anthrax attacks argue that funding these plans amounts to a large waste of taxpayers money. Other, less severe critics suggest that the cost of these plans should be borne entirely by the postal consumer who will benefit from the changes. However, the Postmaster General counters that these are actually costs associated with homeland security and should be funded by taxpayers. Furthermore, passing on these high costs directly to ratepayers may hasten the financial collapse of the Postal Service, which is already on the Comptroller General's High Risk List.

Useful Links

USPS Mail Security Page [http://www.usps.gov/news/2001/press/serviceupdates.htm]

Centers for Disease Control and Prevention Bioterrorism Page [http://www.bt.cdc.gov/]

Federal Bureau of Investigation: Anthrax Investigation Page [http://www.fbi/gov/majcases/anthrax/amerithraxlins.htm]

GAO Report Benefits and Risks of Food Irradiation (GAO/RCED-00-217) [http://www.gao.gov/archive/2000/rc00217.pdf]

GAO Report Diffuse Security Threats: Technologies for Mail Sanitization Exist but Challenges Remain (GAO-02-365) [http://www.gao.gov/new.items/d02365.pdf]

In: Focus on Bioterrorism
Editor: George T. Foster, pp. 39-42
ISBN 1-60021-185-2

Chapter 4

ANTHRAX IN THE MAIL

Nye Stevens and Frank Gottron

ISSUE DEFINITION

Discovery of anthrax spores in mail arriving at media and congressional offices in October 2001 changed the relationship of the U.S. Postal Service (USPS) to its customers and to Congress. In the immediate aftermath of the discovery, delivery of mail to 137 million addresses each day became a source of fear to some Americans and of uncertainty to many more. No institution was affected more directly than Congress itself, where dozens of staff members tested positive for anthrax exposure and thousands were barred from their offices. The use of mail as an instrument of bioterrorism now has Congress questioning whether the largely independent Postal Service has the capability and resources to be in the front lines of homeland defense. Already in deep financial trouble, USPS is looking to Congress for substantial financial assistance, and a fundamental change in its role in government is back on the agenda.

CURRENT SITUATION

Letters containing anthrax, postmarked as early as September 18, were delivered to the offices of the *Sun* tabloid, the New York *Post*, NBC and CBS News, and Senate Majority Leader Thomas A. Daschle. Somewhat belatedly, tests were undertaken in selected mail distribution facilities, where more spores and another infected letter were found. In all, 284 postal locations have been tested, of which 23 tested positive, though some had as few as one to 10 spores. Twenty-two cases of anthrax, including five deaths, have been confirmed by the *Centers for Disease Control and Prevention*. USPS carries current security *information for customers and employees* on its Web site. The Hart Senate Office Building reopened on January 22, but the Brentwood mail processing facility in Washington remains closed. USPS claims that it has sterilized and made available to agency mail rooms all the government mail that had been held since October. Some mail from that period is still trickling in to congressional and executive branch recipients, as internal mail distribution continues to struggle with its backlog. Some government offices, including the Library of Congress, have urged correspondents to use alternatives to the mail, since delivery is unreliable.

USPS has been struggling financially since the attack. Even though mailers agree to a 7.7% rate increase effective June 30, USPS faces a deficit for the year of $1.5 billion, its third in a row. After three quarters of fiscal year 2002, mail volume is down 5.7%. USPS blames most of this on the slow economy rather than on fear of contamination. Others fear the loss may be permanent, driven by the ready availability of alternatives and three rate increases in two years.

Prior to the discovery of anthrax-tainted mail, no procedure was in place to sterilize potentially harmful biological agents in the mail. The interim solution consists of sending select mail, including all mail destined for the federal government offices in the Washington, D.C. metropolitan area, to a sterilization facility in New Jersey for irradiation. This processing adds about four days to delivery time for a typical piece of mail. Mail will continue to be trucked to these facilities until the USPS can construct its own sterilizing facility in the D.C. area. It is hoped that the new facility may reduce the delay to two days.

POLICY ANALYSIS

Even before the September attacks and the anthrax scare, postal insiders had recognized the need for change. A combination of rising costs, declining revenues, lagging productivity growth, and enhanced competition from electronic and package delivery alternatives had created a financial crisis. The board of governors had asked Congress for greater flexibility in controlling its costs (which are 80% labor) and its rates, which can be changed only after a 10-month regulatory review in which USPS's competitors are highly influential. The General Accounting Office put USPS on its high-risk list, calling its current path *unsustainable*.

The anthrax attacks have brought new and unanticipated pressures on USPS to step up its law enforcement activities and its investment in such costly new technology as irradiation, neither of which is related to the USPS mandate to deliver the mail and break even on costs. USPS has estimated the cost of sanitation hardware and systems, training, testing, and new security procedures at $3 billion to $4 billion. It is difficult to see how USPS can impose such expenses on mailers; each rate increase further enhances the attractiveness of e-mail, telephonic billpaying, and fax transmission as cost-effective alternatives. The problem is compounded by a dropoff in revenues from declining use of the mail. In a statement on November 8 before the Senate Appropriations Subcommittee on Treasury and General Government, the postmaster general said that USPS needs assistance to cover both categories of consequences from the attacks: $3 billion to $4 billion in added costs and $2 billion in lost revenues.

USPS received $175 million from the White House Emergency Response Fund on October 24. Most of this has been spent for an interim solution to sanitize certain mail and other immediate emergency response equipment and medical measures. *P.L. 107-117*, the Defense Appropriations Act, provided another $500 million for emergency expenses, with the condition that USPS provide a detailed emergency preparedness plan before the money could be disbursed. USPS released its *plan* in early March. It set forth a need for $87 million in additional appropriations for mail security in FY2002, $800 million in FY2003, and $898 million in FY2004. Nothing was requested in the President's FY2003 budget for mail security, but USPS apparently expects that its request will be considered as a supplemental in the context of homeland security. *H.R. 4775*, the FY2002 Supplemental Appropriations Bill for Further Recovery From and Response To Terrorist Attacks, does provide the $87 million that the USPS plan says it needs in FY2002.

OPTIONS AND IMPLICATIONS FOR U.S. POLICY

The anthrax crisis, if it serves to remind Americans how dependent they are on dependable and affordable mail service, may help attract attention to a long-term solution to postal financial problems. *H.R. 4970*, the Postal Accountability and Reform Act, proposed broad changes in the USPS governing statute designed to increase USPS competitiveness. In part because there was no assurance of floor action during this Congress, *H.R. 4970* was not reported from the Government Reform Committee in a markup on June 20, 2002.

Privatization was recently a widely discussed option for the Postal Service, following in the steps of many foreign countries. That seems less likely in view of the increasing law enforcement and public health responsibilities being thrust upon USPS.

USPS is planning a multilayered semi-redundant system for protecting postal workers and mail recipients. The proposed system is built on risk reduction, threat detection, investigation, and intervention. Risk reduction includes procedural changes such as switching from pressurized air to vacuuming to remove dust from sorting machines, HEPA filters, and worker education. Threat detection is likely to be based on a real-time PCR-based detector that should be able to detect even very small amounts of viral and bacterial bioterror agents. To ease investigation of terrorist attacks, USPS is developing tools to make mail easier to trace back to individuals. Intervention in the short term consists of systematic sterilization of mail destined to high probability targets using irradiation. This may be expanded in the future to include all non-commercial mail (not before FY2005).

USPS chose electron beam, or e-beam, technology to sterilize most mail. E-beams, however, do not penetrate very deep and are therefore unsuited for sterilizing packages. USPS is working on an X-ray-based sterilization system for use on packages. Some have raised concerns that these irradiation based methods of sterilization may damage mail. For example, paper and plastics can be embrittled and discolored; electronics can be damaged; unexposed film exposed; gems, medical samples and seeds could be destroyed. More disturbingly, some congressional staff have complained of nausea, skin irritation, and bleeding from their nose and ears after handling irradiated mail.

The Senate Sergeant-at-Arms has formed the Legislative Mail Task Force to determine if these symptoms are caused by the irradiated mail. The task force includes experts from USPS, the Department of Defense, the National Institute for Occupational Safety and Health (NIOSH), the Office of the Attending Physician, and the General Services Administration. In April 2002, the task force released the NIOSH report, which stated "We did not find evidence suggesting the potential for long-term health effects from handling of irradiated mail." USPS also believes irradiated mail is safe. However, a report released July 2, 2002, by the Office of Compliance (OoC) does not completely support the NIOSH findings. OoC suggested that further monitoring and additional studies should be done to rule-out any long-term effects caused by handling irradiated mail.

ROLE OF CONGRESS/LEGISLATION

Congress faces a series of difficult decisions: how much should be provided to USPS for further mail sanitation and process changes; is the dropoff in volume and revenue a reason for added appropriations; how much can be expected from postal ratepayers already facing a third rate increase since 2000; should the USPS debt limit be raised; and what changes are needed in the USPS relationship to the rest of government, including a new Department of Homeland Security if it is created?

In: Focus on Bioterrorism
Editor: George T. Foster, pp. 43-66
ISBN 1-60021-185-2

Chapter 5

THE HISTORY OF BIOTERRORISM

Teri E. Lassiter* and Michael Conte
University of Medicine and Dentistry of New Jersey, New Jersey Dental School, Newark, NJ

ABSTRACT

On September 11, 2001 the lives of many American citizens changed in an unexpected way. The acts of terrorism that took place on that date caused us to become more conscious of our surroundings. Terrorism was brought into our lives in a different way, we were not watching it unfold on the news in another country, but were watching it unfold in the United States. Foreign terrorists had become a part of our society, they had moved into our neighborhoods and had become the ones that succeeded in destroying our sense of security. The acts of terrorism that took place on September 11 caused us to face terrorism once again at home. With the concern of terrorism still on the minds of many the concern soon changed to bioterrorism when letters containing anthrax began to appear in the mail, closing two major postal facilities on the east coast. On October 4, 2001 a Florida man was diagnosed with anthrax, he died on October 5, 2001 making him the first person in the United States to die as a result of a bioterrorist act.

The use of biological weapons has been attempted throughout history as a means to terrorize the enemy during wartime or as a public threat during times of peace. The use of biological pathogens has been well documented with bacteria, viruses, fungi or toxins, being used on poison arrows or to bomb an area with deadly bacteria and can be traced to as early as 190 BC by Hannibal who hurled venomous snakes onto the enemy ships of Pergamus at Eurymedon to panic and injure enemy sailors. The

* Correspondence concerning this article should be addressed to:
Teri E. Lassiter, MPH University of Medicine and Dentistry of New Jersey – New Jersey Dental School Manager, Office of Infection Control and Environmental Safety, 110 Bergen Street PO Box 1709, Newark, NJ 07101-1709;

concern regarding the use of biological weapons remains throughout the United States and in foreign countries.

This chapter will focus on the historical aspects of bioterrorism tracing their use over the centuries to the current day. It will include a discussion of events that took place during World War I and World War II, related to the use of biological weapons and the decision made by President Nixon to discontinue offensive biological research and destroy the United States' arsenal.

Keywords: bioterrorism, smallpox, anthrax

OVERVIEW

On September 11, 2001 the lives of many people changed when terrorists attacked the World Trade Center and the Pentagon, killing thousands and disrupting the lives of millions of people. We sat and watched as the towers fell, as a portion of the Pentagon went up in flames, while another section of the building collapsed, and as a plane was forced down in the countryside of Pennsylvania. The change was unexpected and we were unprepared. The acts of terrorism that took place in September 11, 2001 caused us to become more conscious of our surroundings and realize that the United States was vulnerable to the possibility of another attack. Following the terrorist attacks of September 11, the fall of 2001 brought more anxiety as anthrax letters began to appear in the mail causing two major postal facilities to close and the subsequent deaths of five people. The threat of bioterrorism became a major concern, not just to the American government but also to the American people as they wondered what was next.

The threat of bioterrorism was now real in the minds of many Americans and others throughout the world while the possibility of another attack loomed in the minds of policy analysts, academicians, and journalists. The utilization of biological pathogens had been well documented throughout history with bacteria, fungi, viruses or toxins being employed to make poison arrows, contaminate well water, or in the making of bombs. Events related to the use of bioterrorism or biowarfare have been traced as far back as Biblical times, but in the United States programs related to germ warfare were ended under President Nixon. As we have entered the 21st Century bioterrorism has become a topic that is debated in the common press and is a concern to the general public. With the collapse of the Soviet Union in 1991, it is unclear what happened to microbe stocks and technology developed by Soviet scientists, as well as the scientists. Some have hypothesized that they have been sold on the black market, while others think that the scientists have been lured to other countries to aid in their development of biological weapons to be used against the United States and other countries that are not sympathetic to their religious or political causes.

The term bioterrorism evolved from biowarfare that was used to describe earlier efforts in the dispersal of biological weapons. While there is no commonly accepted definition of the term bioterrorism, governmental agencies, journalists and other groups provide a series of definitions that focus on the idea that bioterrorism is intended to cause a major disruption of

Michael Conte, DMD, MPH University of Medicine and Dentistry of New Jersey – New Jersey Dental School Assistant Dean of Clinical Affairs, 110 Bergen Street PO Box 1709 Newark, NJ 07101-1709.

political and societal systems. Although there is no one accepted definition of bioterrorism the Center for Counterproliferation Research at the National Defense University defines bioterrorism as the assumption "to involve the threat or use of biological agents by individuals or groups motivated by political, religious, ecological, or other ideological objectives" (Center for Counterproliferation Research, National Defense University, 2001, p. 3). Bioterrorism can also be defined as "the intentional release, or threatened release, of disease producing living organisms or biologically active substances derived from living organisms for the purpose of causing death, illness, incapacity, economic damage, or fear" (Weinstein and Alibek, 2003, p. 2) or as the unlawful or threatened use of microorganisms or the toxins derived from living organisms to produce death or disease in humans, animals or plants. An act of bioterrorism is intended to create fear and/or intimidate the governments or societies in the pursuit of religious, political, or ideological goals that are different from the mainstream.

The use of biological weapons has been recorded for centuries and can be traced to Biblical times. With the procurement and production of biological agents becoming easier as technology improves and globalization permits the exchange of information via the Internet, there are no signs that the threat of bioterrorism will disappear in the near future.

Biological warfare, biological terrorism, biological weapons and bioterrorism are all forms of terrorism designed to intentionally use pathogenic microorganisms and/or toxins against an individual, group of individuals, plants, or animals to cause sickness, diseases, or ultimately loss of life. Biological warfare has been utilized for as long as humans have been at war with each other; infecting communal water supplies with human and/or animal corpses and catapulting victims afflicted the plague over the walls of besieged cities can be accounted for as early as 1346.

Early Use of Biological Weapons

An early reference to the use of anthrax can be found in Exodus 8 and 9 of the Bible. During the fifth and sixth plagues an infectious disease kills the livestock of the Egyptians and then the Egyptian people while sparing the Hebrews and their livestock (Cieslak and Eitzen, 1999; Exodus 8: 20-32, Revised Standard Edition). In I Samuel, plague is released when God turns against the Philistines with rats appearing across the land bringing death and destruction (1 Samuel 4:5-21 and 5:1-12, Revised Standard Edition). The plague has also been associated with the defeat of Athens by Sparta during the Peloponnesian War (431-427 BC); in the crowded city of Athens, as a result of a plan devised by Pericles, the Athenians fell victim to a disease that spread throughout the Mediterranean. With no preventive measures available, it has been estimated that approximately one-third of the population was killed. Scythian archers in 400 BC dipped their arrows in blood, manure or tissue from decomposing bodies to poison their enemies. In the 6th century BC the use of bioweapons was noted on several occasions, including the use of rye ergot by the Assyrians to poison the wells of their enemies and during the siege of Krissa King Solon of Athens poisoned the water supply with hellebore, an herb purgative. During the 2nd century BC in the Battle of Eurymedon, Hannibal's forces hurled earthen pots that were filled with serpents at the decks of ships navigated by King Eumenes of Pergamus forcing a battle between men and snakes; Hannibal's forces won the war.

The use of bioweapons continued into the 14th century AD to wage war and promote terror. The Tartar army fell victim to the plague during the siege of Kaffa (present day Feodosija, Ukraine) between 1346 and 1347. Kaffa was a Genoese cathedral city and the main port for Genoese merchants and their trade industry. Located on the Crimean Peninsula of the Black Sea, Kaffa connected the shipping industry to Tana (Azov, Russia). Tension built along the trading route between Italian traders and their Mongol hosts that lasted until the 1340s when the city flourished and became heavily fortified (Wheelis, 2002). Tartar forces under the leadership of Kipchak khan Janibeg, and reinforced by Venetian troops who were competitors of the Genoese laid siege to Kaffa hoping to remove the Genoese from power and end their dominance of the east-west trade route. With its supplies cut off, the city remained helpless until 1347 when they realized that their enemies we dying at an alarming rate. Janibeg's army had been overcome by the plague that was already raging in China. Before calling off his siege of Kaffa, Janibeg had what was left of his troops catapult the infested corpses of his dead men over the walls of Kaffa and into the city in an effort to infect the Italians (Real Instituto Elcano, 2004). In an effort to save themselves the Italians dumped the bodies into the sea, however, it was too late. The plague caused devastation in Kaffa, those who were able to escape left the city for Italy, however, the plague traveled with them; it has been hypothesized that this was the cause of the medieval plague that swept across Europe killing approximately 25 million people. The Russians used a similar tactic against Sweden in 1710.

With increased travel to the "new world" in the 15th century the Spaniards, French and English found it necessary to conquer the native inhabitants that they came into contact with in North and South America. It is important to note the influence of epidemic infectious diseases on the history of the conquest of America, not simply the use of guns and soldiers. Childhood diseases that were common in Europe were brought from the Old World by the European conquistadors that spread among the native populations. It was through the use of certain diseases as biological weapons that aided in the devastation and depopulation of the native inhabitants as they were slowly conquered by their invaders.

Smallpox was the weapon of choice in two major attacks against the native populations in North and South America. In the early 16th century the Spanish occupied the Caribbean islands of Hispaniola and Cuba and imported slaves from West Africa to work in the mines and plantations, the West Africans came from areas where smallpox was endemic and thus introduced the disease to the native populations of the Americas, spreading from Hispaniola to Cuba and then onto the Mexican mainland. As the Spanish continued to advance into Mexico, the disease was spread as Hernando Cortes conquered the Aztecs and destroyed their empire. The disease continued to spread into Guatemala and then into what is presently Peru wreaking havoc along the way (Tucker, 2001). By 1530 Francisco Pizarro had invaded Peru, bringing with him not only Christianity but also the smallpox virus. It has been reported that those who had not been infected by the disease as a result of earlier exposure where now exposed by Pizarro's troops as they distributed variola-contaminated clothing to the Incans in an effort to continue their conquest.

The most famous early use of smallpox as a biological weapon occurred during the French-Indian War during Pontiac's rebellion in 1763. Smallpox was the bioweapon of choice again by the forces of Sir Jeffrey Amherst in the aftermath of the French and Indian

War (1754-1763) when Britain fought France and its Indian allies for the possession of Canada. As commanding general of the British forces during the war, Amherst was well known for his victories against the French in Britain's quest to acquire Canada, as well as helping to make Britain the world's chief colonizer at the conclusion of the Seven Years War among the colonial powers. While he was well known for his achievements for Britain, he is infamously known for his use of smallpox-laden blankets against the Ottawa Indians in Pennsylvania when he confiscated their land. Pontiac, the Ottawa chief, had sided with the French and wanted to oust the British from their land in Canada and Mississippi and formed a military alliance with other tribes in the area. Pontiac and his forces were successful and were able to defeat the British forces in western Pennsylvania, threatening Fort Pitt. Amherst sent a letter to Colonel Henry Bouquet requesting the use of smallpox against the Indian tribes in order to defeat them; Bouquet was in favor of using this action but expressed concern about the troops becoming infected (Fenn, 2000). Smallpox decimated the Native Americans, who had never been exposed to the disease before and had no immunity, thus the British were able to proclaim victory.

With the germ theory of disease being well established by World War I, Louis Pasteur and Robert Koch had proven that bacteria and viruses could transmit disease and illness. Armed with this knowledge during World War I, German scientists worked with military officials to develop weapons for sabotage during the war generally aimed at livestock that was being shipped to the Allies from neutral countries. The Germans planned to use the stocks of *Bacillus anthracis*, *Pseudomonas mallei*, and a wheat fungus, *Puccinia graminis* against their enemies; with anthrax being used against reindeer in Norway and against Romanian sheep destined for Russia, with a combination of anthrax and glanders against Argentinian livestock that was to be used by the British and Indian armies (Christopher, Cieslak, Pavlin, and Eitzen, 1997; Robertson, 1995).

Since the Japanese had refused to sign the Geneva Protocol in 1925 and ignored the fact that the development and use of biological weapons was banned, they believed that if biological weapons were banned then they should exploit this time to make a greater weapon. During the Japanese-Sino War (1937-1945) and World War II the Japanese formed a biological weapons section of their Imperial Army called Unit 731. Unit 731 was staffed with over three thousand scientists and technicians that were housed in 150 buildings that were located within the unit and at five other sites; it was disguised as an Epidemic Prevention and Water Supply Unit. The Japanese carried out experiments using prisoners of war in the Manchurian state, Russian and American prisoners of war, and on Chinese nationals using *Yersinia pestis* and other diseases. The human test subjects would be infected with the disease under study, the plague, smallpox, botulism, smallpox and other disease, and would be watched to determine the effects of the disease. Vivisections were performed on the infected test subjects removing their organs to see the effects of the diseases. Unit 731 was finally destroyed by Japanese troops in 1945.

The United Kingdom and Biological Warfare

Biological and chemical weapons research in the United Kingdom was established at Defence Science and Technology Laboratory, Porton Down, commonly know simply as

Porton Down. Established in 1916 as the Royal Experimental Station, it was used as a test site for chemical weapons research, specifically chlorine, phosgene and mustard gas for use during World War I. From 1902 through the end of the World War II, research continued on the use of mustard gas in war and in the defense of the country. By 1926 the government added the chemical defense aspects of Air Raid Precautions for the civilian population to the duties of those at Porton Down. After the ratification of the Geneva Protocol the agents that were being used at Porton Down were destroyed and research was scaled down to the hazard assessment of chemical warfare and defensive methods (Carter, 2000; 1992).

As the possibility of biological warfare became more apparent during World War II with threat from the Germans and Japanese, research at Porton Down was stepped up and continued to thrive for twenty years making advances in the field of microbiology although they were not used during wartimes (Carter, 1992). One of the most well known experiments carried out by the scientists at Porton Down took place on Gruinard Island.

Gruinard Island is located two miles off the northwest coast of Scotland in Gruinard Bay and is approximately 1.5 square miles in size. In 1942, British biologists chose the small island as the test site for the development of the first anthrax bomb during World War II amid fears that the Germans or the Japanese may attack the United Kingdom with biological or chemical weapons.

Prior to being used a testing ground for the government, Gruinard Island was home to crofters and their sheep, and was used by people from the mainland as a picnic area. By 1941, the island housed one shepherd and his flock of sheep that belonged to a butcher on the mainland, it was also in 1941 that the British ministry of Defense took control of the island, declaring it a prohibited area and forcing the shepherd to leave the island for the mainland.

Led by Sir Oliver Graham Sutton, a research team of fifty was chosen to perform the tests on Gruinard Island. Stationed on the mainland this group prepared the anthrax slurry for the tests that would take place through 1943. Approximately one hundred and fifty-five sheep were purchased from local villagers for the experiments, with fifteen slated for the first test in July of 1942. Once taken to the island, the sheep were placed in individual wooden crates with their heads outside and the sides of the crates covered to prevent the sheep from ingesting the anthrax by licking their wool or the outside of the crates. The sheep were then placed one hundred yards from the point where the anthrax bomb would be detonated in an arc shaped formation. The first bomb was set for detonation and was exploded four feet from the ground. After the detonation, the sheep's heads were covered with canvas to prevent the inhalation of stray anthrax spores, it is to be noted that the scientists that placed the coverings were wearing protective equipment to prevent their exposure to anthrax, the sheep were then moved to a holding area and tethered so that the effects of exposure could be studied. On the seventh day following the release, death started among the sheep. Of the original fifteen sheep only two survived, they were furthest from the detonation site. Additional tests were run in July 1942 that proved that anthrax could be lethal at a distance of up to four hundred yards (Guillemin, 2005). It was determined from these tests that if 100 kg of anthrax were sprayed on a major city, that more than 3 million people could be killed. The tests on Gruinard Island were considered successful, however, a problem remained – the island had to be decontaminated.

During the time that the experiments were performed anthrax spores were able to penetrate the soil that primarily consisted of sandstone and a topsoil layer of peat bog. First attempts were made in 1943 by burning all of the surface vegetation that consisted of heather, moss, and braken; this was unsuccessful, forcing the military to keep people off of the island. Several other unsuccessful attempts were made to decontaminate the island until 1986, when the final decontaminated effort took place. In 1986, two hundred and eight tons of formaldehyde solution diluted in seawater was sprayed over the entire island, and the topsoil that was located at the dispersal site was removed. In order to determine if the decontamination effort was successful, a herd of sheep was placed on the island and their reactions were monitored. The sheep remained healthy and the decontamination effort was considered a success; in April of 1990, forty-eight years after the island was taken over for experimental purposes, it was deemed safe for public use. However, the island remains uninhabited.

At the end of World War II, those in charge of the British program followed the lead of the United States and continued their biowarfare program into peacetime, as well as their atomic warfare program. The Royal Air Force (RAF), under the name of the Microbiological Research Department, pursued research into the biology pathogens, including Brucellosis, tularemia, and the plague, that were responsible for other diseases. Britain biological weapons research continued through the 1960s and in to the 1970s when they signed the Convention on the Prohibition of the Development, Production and Stockpiling of Bacteriological (Biological) and Toxin Weapons and on their Destruction commonly known as the Biological and Toxin Weapons Convention. Once the Convention was signed, the RAF renamed its research program at Porton Down the Microbiological Research Establishment and reviewed its program until it was restructured in 1979 to focus on defensive research. Today Porton Down is known as the Defence Science and Technology Laboratory; it is the center of scientific excellence for the Ministry of Defence in the United Kingdom and houses one of the largest groups of scientists and engineers in public service in the country.

United States Program Development

The United States began research into the offensive use of biological weapons in 1942 under President Theodore Roosevelt at Camp Detrick, now known as Fort Detrick and is the home of the United States Army Medical Research Institute of Infectious Diseases (USAMRIID). Fort Detrick is located in Frederick, Maryland and lies approximately fifty miles north of Washington, DC in the heart of Frederick County. Camp Detrick was initially established as a training field for the encampment of 104th Aero Squadron of the 29th Division Maryland National Guard in 1931 and named Detrick Field in honor of the late Dr. Frederick L. Detrick, flight surgeon of the unit (Covert, 2000).

The program was called the War Reserve Service with testing sites located in Mississippi and Utah and a production facility in Indiana. In the early years of this offensive program experiments were conducted using *Bacillus anthracis* (anthrax), botulinum toxin (botulism), *Yersinia pestis* (plague), *Francisella tularensis* (tularemia), *Coxiella burnetii* (Q fever), *Brucella suis* (Brucellosis), staphylococcal Enterotxin B, and Venezuelan equine encephalitis (VEE) virus (Covert, 2000).

Building 527, the 8-ball Building at Fort Detrick, was completed in 1950 and was home to Project Whitecoat in 1952. The 8-ball was a four-story steel sphere that was built for the explosions of munitions. Experiments to test exposure to biological agents between 1954 and

1973, allowed American citizens to enlist in the army as conscientious objectors in a non-combative status. Over two-thousand male Seventh-Day Adventists volunteered to be human guinea pigs and allowed themselves to be exposed to biological agents and observed later for effects during this period, mainly for the study of diseases and their cures. Scientists exploded germ weapons inside of the 8-ball that created mists of infectious aerosols (Guillemin, 2005; Miller, Engelberg, and Broad, 2001) that the volunteers were exposed to by breathing through masks attached to chamber ports; after exposure they were quarantined for observation and medical treatment, if necessary. Recruitment for Project Whitecoat ended in 1973, and concluded in 1975, because of incomplete data recording mortality rates could not be accessed (Pittman, Norris, Coonan, and McKee, 2005). Building 527 burned down in 1975, and the 8-ball structure remains, it has been placed on the National Registry of Historic Places.

As a part of the nation's Cold War defense against the continuing threat of biological warfare, the US Army completed building 470, located within the confines of Fort Detrick, in 1953 costing $1.3 million; the building was used as a pilot plant for the production of bacterial agents as potential warfare agents, including *Bacillus anthracis*, *Francisella tularensis* and *Brucella suis*. Building 470, a seven-story, windowless building, was known as the tower of doom. Through the mid-1960s Army scientists used building 470 to produce *Bacillus anthracis* that was used for research purposes until the program was ended in 1969. The Army took samples from the building in 1970-1971 during its decommissioning and found no traces of anthrax; two rounds of gas decontamination using formaldehyde and steam sterilization followed this. Areas that could not be reached by this process were thoroughly decontaminated separately to insure that nothing remained. Building 470 was turned over to the National Institute of Health for cancer research in 1988; it was vacated and decontaminated for use as a storage facility in 1971. Demolition of building 470 was completed in 2004, prior to which earlier decontamination was reverified by taking additional samples for anthrax.

Research activities at Fort Detrick ceased on November 25, 1969, at the height of the Cold War, when President Richard M. Nixon ended all bioweapons research in the United States. His renunciation of the program came in the wake of the Vietnam War where the United States had used chemical weapons. The use of chemical weapons had been banned in National Security Memorandum 35, dated November 25, 1969, it stated that the President had decided

With respect to Bacteriological/Biological programs:

a. The United States will renounce the use of lethal methods of bacteriological/ biological warfare.
b. The United States will similarly renounce the use of all other methods of bacteriological/biological warfare (for example, incapacitating agents).
c. The United States bacteriological/biological programs will be confined to research and development for defensive purposes (immunization, safety measures, et cetera).
d. The Secretary of Defense will submit recommendations about the disposal of existing stocks of bacteriological/biological weapons.
e. The United States shall associates itself with the principles and objectives of the Draft Convention Prohibiting the Use of Biological Methods of Warfare presented by

the United Kingdom at the Eighteen-Nation Disarmament Conference in Geneva, on 26 August 1969 (Kissinger, 1969).

President Nixon, in his Talking Points, also dated November 25, 1969, noted that by associating the United States with the United Kingdom at the Eighteen-Nation Disarmament Conference that

> This program not only will assure the strength of our defenses but it also will benefit all our people through research into better ways to control and prevent the spread of communicable disease (Nixon, 1969).

In December Nixon extended his renunciation of germ warfare to include biological toxins, thus ending over thirty years of work at Fort Detrick. By 1973 the complete United States arsenal was destroyed; for research purposes the seed stocks were maintained.

The renunciation of biological weapons by President Nixon lead to the ratification of the Geneva Protocol by the United States and support of the Eighteen-nation Disarmament Committee's Convention for the Prohibition of Microbiological Methods of Warfare that would supplement but not supercede the Geneva Protocol. The Geneva Protocol for the Prohibition of the Use in War of Asphyxiating, Poisonous or Other Gases, and of Bacteriological Methods of Warfare of 1925 was enacted to prohibit the use of poisonous gases and bacteriological methods during times of warfare; signed on June 7, 1925, it was entered into force on February 8, 1928. The Protocol was enacted to prohibit the use of chemical and biological weapons, but said nothing about their production, storage, or transfer and only affected those that were parties to the accord, thus there were no restrictions placed on non-parties except those of customary international law. It has also been debated as to whether the Protocol's banning of bacteriological agent included a ban on everything that is considered a microbial, including viruses.

The Convention on the Prohibition of the Development, Production and Stockpiling of Bacteriological (Biological) and Toxin Weapons and on their Destruction commonly known as the Biological and Toxin Weapons Convention (BTWC) was signed in 1972 simultaneously in London, Moscow, and Washington and entered into force on March 26, 1975. The BTWC was the first multilateral disarmament treaty that banned an entire category of weapons, with the exception of medical uses. This document was adopted to prevent the parties of the convention from developing, producing, stockpiling, or acquiring "biological agents or toxins 'of type and in quantities that have no justification for prophylactic, protective, or other peaceful purposes,' as well as related weapons and means of delivery" (Federation of American Scientists, n.d.). For President Nixon, the BTWC presented an opportunity to reinforce his renunciation of the biological weapons program in the United States, as well as end the possibility of biological agents being used in the Vietnam War (Frischknecht, 2003). The BTWC was seen as an extension of the Geneva Protocol because while the BTWC did not specifically outlaw the use of such weapons, only their production, storage, and means of delivery, the use could be considered unlawful under a number of preexisting Geneva Convention protocols and international law. On April 10, 1972, Nixon signed the BTWC on behalf of the United States; this officially stopped the expansion of military experiments, the production of agents, and field tests (Guillemin, 2005). The BTWC came into full force in 1975.

In addition to the work being done at Fort Detrick, the United States Army had biological laboratories and testing grounds throughout the country that were used not only for the research and development of bioweapons but also for development of agrochemicals that could be used for crop destruction or deforestation that would deprive enemy troops of ground cover.

It should be noted that even though the former Soviet Union was an original signatory of the BTWC, this did not stop their establishment of the Biopreparat (Chief Directorate for Biological Preparations) in 1973 under the Main Microbiological Industry Agency to perform research and development, and production of offensive biological weapons under the guise of legal and civilian biotechnology research (Alibek, 1999; Davis, 1999). The Soviet Union believed that the United States had not ended their bioweapons program and has simply turned it "into a 'black' weapons program and that the Americans had lied about the disassembly of their program" (Preston, 1998).

The Former Soviet Union

After the Geneva Protocol was augmented in 1972 when the Convention on the Prohibition of the Development, Production and Stockpiling of Bacteriological (Biological) and Toxin Weapons and on their Destruction (BTWC) was enacted and with the Chemical Weapons Convention in 1993, a major incident took place in the Soviet Union, specifically Sverdlovsk, now called Ekaterinburg, and located approximately 850 miles east of Moscow, where anthrax was released from Compound 19. Early reports stated that up to one thousand people had died, TASS, the Soviet news agency, blamed the deaths on a natural occurrence of anthrax among domestic animals, western news agencies said that the Soviet Union had violated the terms of the BTWC (Alibek, 1999).

In March of 1979 a technician at Compound 19 found that a filter needed to be replaced in one of the facility's drying machines and noted this so that the filter would be replaced prior to the next time that the machines were run. These machines were used in the production and fermentation of anthrax; the missing filter was to have been secured over the exhaust pipe to prevent anthrax dust from escaping into the community, without the filter anthrax dust could drift downwind from the plant and expose the city of over one million people to anthrax dust. A filter was clogged and a note was left to replace it the last Friday of March 1979, however, the note went unnoticed by the next shift and the machines were started in order to produced more dry anthrax; several hours into the shift it was noticed that the filter was not in place (Alibek, 1999). Within the next few days, workers at a plant across the street from the facility began to get ill and within a week many had died; others who worked near the plant were soon admitted to hospitals. According to accounts provided by Soviet physicians sixty-four deaths occurred from the 96 cases of human anthrax with onset dates between April 4 and May 18, 1979.

In an analysis by American scientists it was suggested that of the two hundred and fifty identified cases that there were one hundred deaths from inhalational anthrax (Inglesby, et al, 2002). It was not until 1992 that the Soviet government admitted what had taken place and agreed with the charges made by the western press (Leitenberg, 1993). This epidemic represented that largest documented outbreak of human inhalation anthrax in history.

THE ANTHRAX LETTERS – 2001

The first known bioterrorist attack in the United States in the 20th century took place in 1984. The followers of an Indian guru, Gagwan Shree Rajneesh, placed *Salmonella* in salad bars in the Oregon town of The Dalles. Their plan was to cause local citizens to become ill and prevent them from voting in a local election where the Rajneeshee cult was seeking to expand their control of land in the area. The Centers for Disease Control and Prevention was called in to investigate an outbreak of food poisoning, but concluded it was coming from natural causes. This attack resulted in seven hundred and fifty-one becoming ill with diarrheal illness and forty-five people being hospitalized. It was not until a year later that it was found that the Rajneeshee cult was responsible for the outbreak. The second known attack took place in 2001.

Much of the information that is known about the human spread of anthrax has been derived from the Sverdlovsk accident that took place in April of 1979. The anthrax letters that were circulated through the United States postal service in the fall of 2001 provided additional information regarding the dispersal of this disease. Prior to 2001, the last case of inhalation anthrax in the United States was in 1976; between 1900 and 1976 only eighteen cases of inhalation anthrax were reported in the United States, with the majority of cases being seen in among a specified group of workers that included tannery, goat hair mill and wool workers which resulted in an 89% case fatality rate. The cases that occurred in 2001, unlike all other reported cases, were a result of a bioterrorist event.

Following the attacks of September 11, 2001 there were eighteen confirmed cases of anthrax, seven cutaneous cases and eleven inhalation cases; there were four suspect cases that were identified between September 22, 2001 and November 12, 2001. What did these cases have in common? They all occurred on the east coast of the United States, between Boca Raton, Florida and Oxford, Connecticut and the method of dissemination placed many others at risk by occupational exposure.

In the fall of 2001 two letters that were contaminated with anthrax were processed through two United States postal facilities, causing the occupational exposure of over 3,000 postal employees. Two postal facilities (Washington, DC and Hamilton, NJ) on two separate dates processed the two letters that started attacks and panicked the public.

The first anthrax death was reported on Friday, October 5, 2001. An employee of American Media, Inc. located in Boca Raton, Florida was diagnosed with inhalation anthrax on October 4, 2001; he was infected after opening his mail, traces of anthrax were found during an investigation on his computer keyboard and in the company's mailroom. Further investigation at American Media, Inc. revealed that a second employee at American Media, Inc. had also been exposed to anthrax, he worked in the mailroom; his diagnosis confirmed that anthrax was indeed in the building and caused the investigation to expand to include employees at United States Postal facilities that may have processed mail for American Media, Inc. In October 2001, a third employee of American Media, Inc. tested positive for anthrax, the investigation into all three cases is now a criminal investigation; the anthrax cannot be traced to the 9/11 attacks. Testing of postal employees did not reveal any additional cases. The American Media, Inc. building closed and was not declared safe until 2004; American Media, Inc. relocated their headquarters and sold the building.

An anthrax letter was delivered to the NBC News in New York City in October of 2004 and was opened by Tom Brokaw's assistant who subsequently developed cutaneous anthrax. Unlike the mail associated with the cases in Florida, the letter addressed to Mr. Brokaw was found and upon testing contained anthrax. The letter was postmarked on September 18 from Trenton, New Jersey and provided the first information of the source and when the anthrax attacks may have started. The health department of New York City and the Centers for Disease Control and Prevention recommended that other personnel at NBC be tested for anthrax. By the end of September, four other cases had been identified in New York City; one case involved a six-month old infant that was the son of an ABC World News Tonight producer, the other two cases involved employees at CBS news and the *New York Post.* The letter that was received by the *New York Post* was postmarked on September 12 from Trenton, New Jersey and had similar handwriting to the letter sent to NBC.

By October of 2001 the anthrax attacks had spread to the nation's capital. On October 15, an aide to Senator Tom Daschle of South Dakota opened a letter containing a white powdery substance; with the events that had taken place in New York and Florida, a red flag went up and an immediate investigation ensued. The substance was identified and approximately six hundred employees in the Hart Senate Office Building were tested and received prophylactic antibiotics. This was the first letter that was not sent to a media organization, but did originate from Trenton and was postmarked October 9; this letter, unlike the others, contained a postmark from Franklin Park, New Jersey. The letter was tested at the United States Army Medical Research Institute of Infectious Disease (USAMRIID) located at Fort Detrick, Maryland; it tested positive for anthrax. In November another letter was found that had been addressed to Senator Patrick Leahy that had been misrouted because of an error in the zip code, however, it was postmarked on October 9 from the Hamilton, New Jersey mail facility. The Capital and the House and Senate Office Buildings were closed and tested for anthrax, spores were found in the Hart Senate Office Building's mailroom and the building was closed until January of 2002 after it had been decontaminated.

Even though the letters originated at postal facilities, no postal employees were diagnosed with anthrax until October 18 when a letter carrier in Trenton was diagnosed with cutaneous anthrax. Environmental sampling had begun in the postal facilities in New Jersey and Washington, DC. By mid-October two employees from the Brentwood postal facility in Washington succumbed to anthrax, both from inhalation anthrax. Their deaths led to the closing of the Brentwood postal facility. An additional two cases of inhalation anthrax were identified at the Brentwood facility. Five infections were identified at the New Jersey postal facility, two inhalation cases and three cutaneous cases; all five survived.

The last two cases were identified in the Bronx, New York and Oxford, Connecticut. These cases involved two women who were not employed by the federal government or any of those targeted by the letters and they did not work for the United States Post Office. The case in the Bronx was a hospital worker who died of inhalation anthrax in late October and the second victim was a ninety-four year old woman that lived in Oxford, Connecticut and died in November. These two cases were extensively investigated; anthrax spores were found in the post office that delivered the mail to the Connecticut case, but no spores were found in the homes of either case. It was believed that cross-contamination occurred with the spores being attached to the mail as it was processed through the automated machines at the post office.

The Brentwood facility reopened in December of 2003 and the Hamilton facility reopened in March of 2005, both after extensive and costly decontamination. The anthrax attack caused significant disruption to the American government and the American Public. The person who mailed the letters remains unknown, although several suspects have been questioned since 2001. Of the twenty-two individuals infected (See Table 1), eleven were infected with inhalation anthrax and eleven were infected with cutaneous anthrax; five of the individuals died of inhalation anthrax.

Table 1: Anthrax outbreak – 2001 Distribution of Cases by age, gender, demographics and clinical presentation.

Case	Age	Gender[1]	Onset Date	Date of Laboratory Diagnosis	Location of Exposure[2]	Type	Status	Source	Outcome
1	31	F	09/22/01	10/19/01	NYC	Cutaneous	Suspected	Sept. 18 letter	Alive
2	38	F	09/25/01	10/12/01	NYC	Cutaneous	Confirmed	Sept. 18 letter	Alive
3	39	M	09/26/01	10/18/01	NJ	Cutaneous	Suspected	Sept. 18 letter	Alive
4	45	F	09/28/01	10/18/01	NJ	Cutaneous	Confirmed	Sept. 18 letter	Alive
5	63	M	09/30/01	10/04/01	FL	Inhalational	Confirmed	Unfound mail	Dead
6	23	F	09/28/01	10/12/01	NYC	Cutaneous	Suspected	Sept. 18 letter	Alive
7	73	M	09/28/01	10/15/05	FL	Inhalational	Confirmed	Unfound mail	Alive
8	0.6	M	09/29/01	10/15/01	NYC	Cutaneous	Confirmed	Sept. 18 letter	Alive
9	27	F	10/01/01	10/18/01	NYC	Cutaneous	Confirmed	Unfound mail	Alive
10	56	F	10/14/01	10/28/01	NJ	Inhalational	Confirmed	Oct. 9 letter	Alive
11	35	M	10/14/01	10/19/01	NJ	Cutaneous	Confirmed	Oct. 9 letter	Alive
12	43	F	10/15/01	10/29/01	NJ	Inhalational	Confirmed	Oct. 9 letter	Alive
13	56	M	10/16/01	10/21/01	DC	Inhalational	Confirmed	Oct. 9 letter	Alive
14	55	M	10/16/01	10/23/01	DC	Inhalational	Confirmed	Oct. 9 letter	Dead
15	47	M	10/16/01	10/26/01	DC	Inhalational	Confirmed	Oct. 9 letter	Dead
16	56	M	10/16/01	10/22/01	DC	Inhalational	Confirmed	Oct. 9 letter	Alive
17	51	F	10/17/01	10/29/01	NJ	Cutaneous	Confirmed	Oct. 9 letter	Alive
18	34	M	10/19/01	10/22/01	NYC	Cutaneous	Suspected	Sept. 18 letter	Alive
19	59	M	10/22/01	10/25/01	DC	Inhalational	Confirmed	Oct. 9 letter	Alive
20	38	M	10/23/01	10/28/01	NYC	Cutaneous	Confirmed	Sept. 18 letter	Alive
21	61	F	10/25/01	10/30/01	NYC	Inhalational	Confirmed	Unknown	Dead
22	94	F	11/14/01	11/21/01	CT	Inhalational	Confirmed	Oct. 9 letter	Dead

[1] M: male; F: female.

[3] NYC: New York City; NJ: New Jersey; FL: Florida; DC: District of Columbia; CT: Connecticut.

Source: Jernigan, et al., 2002; Jernigan, et al., 2001.

POSSIBLE AGENTS FOR USE IN A BIOTERRORIST ATTACK

The proverbial arms race of yesterday may in fact be the bug race of today. Some world powers admit to such practices while the denial by others will remain in debate. Regardless of the situation, the concern of the possible use of biological agents remains in the forefront of our minds. With the collapse of the Soviet Union and continued unrest in the Middle East and parts of Africa, there is a question regarding who has certain biological stocks and who holds the technology to develop and use the biological agents. It is known that some biological agents have been weaponized by the former Soviet Union and that research continued on the engineering of more virulent strains of various diseases until the fall of the Soviet Union.

Table 2: Characteristics of Selected Biological Weapons Agents.

Disease	Agent	Pathogen Group	Rate of Action	Syndrome	Methods of Dissemination	Immediate Public Health and Infection Control Actions
Anthrax	Bacillus anthracis	Bacteria	Incubation: 1 to 6 days Length of Illness: 1 to 2 days Extremely high mortality	Acute respiratory distress with fever	Inhalation Cutaneous Ingestion	Notify the FBI. Where appropriate notify local and state health departments. Provide public health management and follow-up as needed. Quarantine is not necessary.
Botulism	*Clostridium botulinum* toxin	Toxin	Time to effect: 24 to 36 hours Length of illness: 24 to 72 hours 65% mortality rate	Neurological	Ingestion Aerosol	Report immediately to appropriate authorities.* Use standard precautions. No isolation is required. All foods suspected of contamination should be promptly removed from potential consumers and submitted to public health authorities for testing.
Plague	*Yersinia pestis*	Bacteria	Incubation: 2 to 10 days Length of illness: 1 to 2 days Variable mortality rate	Acute respiratory distress with fever	Aerosol Vectors	Any suspect case should be reported immediately to the local health department. Depending on extent of dissemination, mass prophylaxis may be considered.
Smallpox	Variola major	Virus	Incubation: 12 days average Length of illness: Several weeks 35% mortality rate in unvaccinated individuals	Acute rash with fever	Aerosol	Immediate telephonic communication with local national health authorities is obligatory. Confine infected patients in hospital rooms with negative pressure. Use standard precautions.
Tularemia	*Francisella tularensis*	Bacteria	Incubation: 1 to 10 days Length of illness: 1 to 3 weeks 30% mortality rate	Influenza-like illness	Aerosol	Notify state or local health department. If not available notify the CDC.
Viral Hemorrhagic Fevers **Ebola** **Marburg** **Lassa** **Machupo**	 Ebolavirus Marburgvirus	 Virus Virus Virus Virus	Incubation: 2 to 21 days	Acute rash with fever	Contact Aerosol	Report to local health authority. Institute immediate strict isolation.

Source: Heyman, D.L., 2004; Arnon, et al, 2001; Henderson, et al, 1999.

Biological agents fall into three categories: viral, bacterial and toxins (See Table 2). The biological agents, if chosen for weaponization will be easy to disseminate, have the potential for dissemination over a large geographic area, their detection after a release will be difficult, and symptoms will occur several day or weeks after the release. In addition, the chosen agents may have a secondary spread, such as smallpox; their use can cause public panic, thus stressing the public health system; and the users will have the ability to protect themselves and escape before the effects of the agent take place. At least seventeen nations are known to have offensive biological weapons programs, including Iraq, Iran, China, Cuba, Libya, the Sudan, Syria, Pakistan and North Korea.

The routes of entry for biological agents are the same as for any endemic microorganism: the pharynx and lungs, the gastrointestinal tract, the eyes, breaks in the skin and absorption through unbroken skin. Biological weapons do in fact have some inherent flaws; some consistent with the more traditional terrorist weapons as well as flaws inconsistent with traditional terrorist weapons. The climate of the region, local weather patterns, including wind direction, temperature, humidity, rain fall, and topography of the land all can and do play a role in the effectiveness and longevity of biological agents. Biological weapons also present issues regarding propagation, storage and handling.

Anthrax

Anthrax, a large aerobic, gram-positive spore-forming bacillus, is caused by Bacillus anthracis. As a naturally occurring zoonotic disease anthrax is most commonly seen among herbivores, which are infected after ingesting spores from the soil (Inglesby, et al, 2002) and is rarely seen in the United States. Because of its rarity in the United States, few physicians have had any experience in detecting or treating this disease, including veterinarians who rarely encounter cases of anthrax in the animal population. Anthrax is seen more frequently in both humans and animals in areas of the former Soviet Union and in developing countries where there are less than optimal public health conditions. It should be noted that contact with infected animals could result in human infection. There are three main forms of anthrax infection – inhalation, cutaneous, and oropharyngeal/gastrointestinal. Naturally occurring anthrax is generally seen in the cutaneous form while naturally occurring inhalational cases are rare. No cases of inhalational anthrax have been identified in the United States since 1976, when a craftsman acquired the disease from imported yarn, this case proved fatal; between 1984 and 1994 there were two-hundred and twenty four cases of cutaneous anthrax reported (Migueles and Tuazon, 2001).

As a biological weapon, aerosolized anthrax has been identified as the "single greatest threat as a mass casualty-producing biological weapon" (Migueles and Tuazon, 2001, p. 248), because of this the World Health Organization and the US Department of Defense have focused attention to the continued development of an anthrax vaccine. It has been determined from the Sverdlovsk anthrax incident in 1979 that the LD_{50} in non-human primates ranges from 2,500 to 760,000 spores (Meselson, et al, 1994; Watson and Keir, 1994), however the US Department of Defense estimates that the LD_{50} for humans ranges from 8,000 to 10,000 spores (Meselson, et al, 1994).

Sometimes called a doomsday weapon, anthrax has been weaponized by several countries, including the United States prior to the termination of its offensive program by President Nixon in 1969. Prior to ending the program, the United States produced over 5,000 bombs filled with Bacillus anthracis at Camp Detrick during World War II. Iraq has acknowledged that it conducted research in the development of anthrax for use during wartime and it has been reported that the Aum Shinrikyo cult has attempted at least eight times to disperse anthrax aerosols throughout Tokyo (Migueles and Tuazon, 2001; Inglesby, et al, 1999). Production of anthrax cannot generally be accomplished without advanced biotechnology, therefore it is unlikely that individuals or small groups will be able to manufacture a lethal dose of aerosolized anthrax. It is likely that groups with substantial funding and contacts, and access to the required materials will be able to produce sufficient quantities of anthrax for a successful attack (Inglesby, et al, 1999).

A vaccine against anthrax has been produced and distributed to members of the US armed forces, those involved with anthrax research, and those that are considered to be at high risk. The vaccine schedule includes subcutaneous injections at 0, 2, and 4 weeks, then at 6, 12, and 18 months, an annual booster injection is recommended if immunity is to be maintained (Centers for Disease Control and Prevention, 2000). In his book entitled *Vaccine A*, Gary Matsumoto (2004) suggests, in explicit detail, the United States historical involvement in using unsuspecting GIs as human lab rats to promote their understanding of biological pathogens.

Smallpox

Smallpox is an acute sometimes fatal, contagious disease caused by Variola virus, a member of the orthopoxvirus family and was one of the world's most feared diseases until it was eradicated by a collaborative global vaccination program led by the World Health Organization. The last known natural case was in Somalia in 1977. Smallpox was declared eradicated by the World Health Organization in 1980; it was the first major disease to be vanished from the earth. After the disease was eliminated from the world, routine vaccination against smallpox among the general public was stopped because it was no longer necessary for prevention.

There is growing concern regarding the eradication of smallpox in 1980 and the fact that many since then have not been vaccinated worldwide. This is compounded with the vaccination protection lasting 3-5 years with residual protection diminishing at an unknown rate. The level of immunity, if any, among persons who were vaccinated before 1972 is uncertain. According to a study by Eichner (2003), the amount of residual protection offered by vaccination is dependent on the time elapsed since vaccination. Even 70 years after primary vaccination, 77.6 % of cases still exhibited residual protection. In this study Eichner concluded that age-dependent risks and residual protection have been neglected in current discussions about vaccination strategies and the mathematical models used to predict what might happen should smallpox be released as a biological weapon.

Like Yersinia pestis, it is classified as a category "A" virus by the Centers for Disease Control and Prevention because of its extremely high human-to human transmissibility, morbidity, and mortality (Greenfield, 2002; McClain, 1997). The history of smallpox can be traced to ancient times where it has been found on the sarcophagi of Egyptian mummies with suspicious eruptions of the skin. These findings correlated with early texts that recounted outbreaks of diseases (Eyler, 2003; McClain, 1997). Smallpox is spread easily from person-to-person, usually through the respiratory tract (droplet spread) or skin inoculation (Heymann, 2004) and can be isolated from all body fluids, resulting in the need to isolate and quarantine its victims. The virus can also be spread through contaminated clothing or bed linens and poses a significant concern to health care providers.

The use of smallpox as a bioweapon can be traced to as early as the 15th century when Francisco Pizarro, a Spanish soldier of fortune, accompanied by 168 soldiers entered the Inca City of Cajamarca where it is believed that they distributed contaminated clothing to the natives (Diamond, 1999). It has been questioned if this account is accurate since smallpox had been introduced to South America earlier in the 15th century by the Spanish as a means to aid in their subjugation of Honduras, Guatemala, and Nicaragua (Patterson and Runge, 2002). Between 1754 and 1767, during the French and Indian War, Sir Jeffery Amherst gave Indians who were loyal to the French, blankets laden with smallpox in order to take command of Fort Carillon (Eitzen and Takafuji, 1997). In June of 1763, Amherst enlisted the assistance of one of his officers suggesting that smallpox be used as a means to destroy the coalition of Native American tribes. In a letter written to Amherst from Colonel Henry Bouquet, he suggests the use of smallpox laden blankets. Blankets and handkerchiefs that had been used by American soldiers in the infirmary were acquired and consequently distributed to the Native Americans subsequently causing an outbreak of smallpox (Wheelis, N.D.). During the American Civil War allegations were made against the British Colonial Commanders, stating they considered distributing smallpox-laden blankets to the Native Americans, since it had been successful during the French-Indian War. During World II, the Japanese forces explored the use of smallpox as a bioweapon, at Unit 731 (McClain, 1997).

Plague

The plague has been responsible for countless epidemics as well as 3 pandemics throughout history and can be traced as far back as 1320 BC where it was interpreted as being bubonic plague in the Book of Samuel I. The plague was once thought to be a disease associated with religious beliefs of persecutions and horror with the disfigured bodies of plague victims reflecting bizarre happenings in the physical world (Yom, 1997). The first recorded plague pandemic occurred about 541 AD and killed as many as 100 million Europeans. The second pandemic was also known as the black death or great pestilence because of the blackening of the skin just prior to death; beginning in the mid-12th century and lasting for over six years, it has been estimated that 24 million people died from the plague, with another 20 million dying by the end of the 14th century. The third or modern pandemic occurred in China in 1894 and was spread throughout the world by modern transportation. It was also in 1894 that Alexandre J.E. Yersin, for whom the bacteria was named, discovered that Yersinia pestis satisfied Koch's postulates for bubonic plague (McGovern and Friedlander, 1997).

Yersinia pestis, a gram-negative bacillus, causes the plague; infected fleas are the most common vector for the spread of this disease. Animal to human transmission occurs typically from fleabites where the flea's previous host was an infected animal, contact with animal pelts containing infected fleas and contact with infected animal body fluids while field dressing the animal. Human to human transmission of plague occurs also from fleabites where the fleas previous host was an infected human and contact with infected human body fluids more specifically droplets generated by coughing (Persell, et al, 2002).

The Centers for Disease Control and Prevention have categorized the plague as a category "A" agent; as such it is considered a high priority because it poses the greatest possible threat on public health resources, may spread across a large area, create a need for public awareness, and a great deal of planning to protect the public's health. If used in an aerosol attack Yersinia pestis could cause cases of the pneumonic form of plague. Once the disease is acquired, the bacteria can spread from person-to-person, especially to those that have direct, prolonged exposure to the infected person. The delay between exposure to the bacteria and becoming symptomatic can allow the infected person to travel, thus exposing others to the disease.

It is believed that the first conscious attempt to utilize the plague as a biological weapon occurred in 1346 when the leader of the Muslim Tatars gave the order to catapult the corpses of their own soldiers who fell victim to the plague, toward the Christian Genoese sailors; the sailors became sick and fled the battle. It is assumed today that the sailors were infected by the endemic plague, not by the infected corpses because entomology suggests that the fleas leave the host once it cools down (Greenfield, 2002).

In World War II, the Japanese developed clay bombs to drop plague infected fleas on their enemy. Fleas not encased in clay dropped from a high altitude would have too wide of a dispersion and basically miss their intended target. Fleas not encased in clay and dropped from a low altitude would put pilot and plane at too great a risk (McGovern and Friedlander, 1997). In the 1950's and 1960's techniques to disperse the plague into the air were developed in both the United States and the Soviet Union biological weapons programs, allowing aerosolized Yersinia pestis to be directly inhaled into the lungs resulting in the contagious and most lethal pneumonic plague (Reilly and Deason, 2002).

Tularemia

G. W. McCoy in Tulare County, California discovered tularemia in 1911. It was first described as a plague-like illness in ground squirrels and other rodents and at one time as rabbit fever. By 1914, it was recognized to be a severe and fatal illness in humans as confirmed cases of human disease were reported. In 1921, Edward Francis identified the transmission of the disease via deer flies as a result of their contact with a host with infected blood. It was Edward Francis who named the disease Tularemia and it was the Soviets in 1959 who suggested changing the genus name to Francisella in honor of Edward Francis work that lead to the current understandings of the disease (Dennis, et al, 2001; Evans and Friedlander, 1997.

As an aerosolized biological weapon, Francisella tularensis could result in one or more forms, depending on the level of personal exposure. Tularemia is not transmittable from human-to-human, so healthcare workers are not at risk of contracting the disease when providing treatment. After exposure or suspected exposure, an early treatment of antibiotic therapy is recommended (Dembek, 2003). Diagnosis is confirmed after a positive culture for F. tularensis is found. It can be cultured from blood, ulcers, conjunctival and pharyngeal exudates, sputum and gastric washings.

Vaccines have had limited success in the past. A whole cell killed vaccine used during an experimental study in the 1930's reduced the frequency of systemic symptoms but did not prevent the formation of lesions. In 1961 another vaccine formula was contrived. This live vaccine strain (LVS) vaccine protected against an aerosol confrontation with virulent Francisella tularensis.

Botulinum Toxin

Clostridium botulinum is an anaerobic, gram-positive spore-forming rod that produces a potent neurotoxin. Its natural habitat is soil from which it can be easily isolated in a laboratory. The species C. botulinum consists of 4 genetically different groups that only have in common the ability to produce botulinum toxin. Seven types of botulism (types A through G) are recognized based on the antigenic specificity of each toxin. The toxin types are distinct by their absence of cross-neutralization. In other words anti-A antitoxin does not neutralize the other toxins B through G (Arnon, et al, 2001). C. botulinum and or its toxin can enter the body via inhalation, ingested, and via wound contamination. Food contamination and aerosolization are to be considered the most likely routes associated with bioterrorism. Food poisoning as a result of botulinum toxin became a problem as the practice of food preservation increased in popularity.

Botulism is not transmitted via human-to-human contact so patient need not be isolated. Once the toxin is absorbed into the human body, it blocks cholinergic synapses resulting in a classic triad:

- Systemic, descending flaccid paralysis progressing to respiratory muscle weakness
- Absence of fever
- Alertness and orientation without sensory defects

A trivalent antitoxin specific to botulism toxins type A, B and E does exist and is administered IV at a dosage of 10 ml of antitoxin diluted 1:10 in 0.9% sodium chloride.

Because of botulinum toxin's extreme toxicity, it was one of the first agents to be considered for use as a biological weapon. During World War II the Germans attempted to develop the botulinum toxin as a weapon but at that time much about its mechanism of action and its lethality in humans as well as animals was not fully understood. At that time botulinum toxin was referred to as Agent X. The United States spent decades trying to develop a botulinum toxin weapon. The most important advancement that resulted from this early biowarfare research was the development of the botulinum vaccine that is still used today (Middlebrook, and Franz, 1997).

Glanders

Glanders is caused by *Burkholderia mallei*, a small Gram-negative bacillus that primarily infects horses, mules and donkeys. It has been recognized since Ancient Greece but humans have seldom acquired the disease either because the ill animals had low concentrations of organisms or because the strains virulent for animals are less virulent to humans (Georgiades, and Fishman, 2001). The disease has also significantly decreased since the replacement of the horse with the automobile.

Glanders has been transmitted to humans via inhalation, contact with oral, nasal and conjunctival mucosa and contact with open skin. It results in severe disease and has a high mortality rate. Since there is no vaccine for glanders, standard precautions should be practiced to prevent further human-to-human transmission. Acute infection via inhalation of B mallei in humans is characterized by necrosis of the tracheobronchial tree, pustular skin lesions and pneumonia. Human contact with *B. mallei* via the skin and open wounds is characterized by multiple abscesses and signs of sepsis (Srinivasan, et al, 2001).

Once a proper diagnosis is reached, treatment with specific antibiotics has been effective. The untreated septic form of Glanders has a 95% mortality rate and a > 50% mortality rate if treated. Mortality rate for treated localized Glanders is 20 percent (Greenfield, 2002).

Aerosolized *B. mallei* isconsidered a highly effective biological weapon because of its lethal and contagious natures. The Germans in World War I experimented with B mallei as a biological weapon. Few cases have occurred in the United States. Those that have occurred are usually associated with occupational exposure such as veterinarians, abattoir workers, equine caretakers and laboratory workers. Outside of the United States cases are more prevalent, and it is considered endemic in South East Asia and the Middle East. It is proposed that one case of Glanders in the United States in an individual not employed in the aforementioned occupations should suggest a biological attack (Greenfield, 2002).

Conclusion

Humans have been using biologic warfare since antiquity. It is believed that during ancient times no one understood the direct relationship between microorganisms and the resulting disease. It was not until 1877 that Robert Koch a physician, born in 1843 Germany, continued with the work of Louis Pasteur a research scientist, born 1822 France. Louis Pasteur used a microscope to look at the microorganisms he found in vats of beer, wine, milk and vinegar. Through his experiments he was able to prove that air contained living organisms and that these microbes can cause putrefaction. He also proved that heating the liquid they were in could kill microbes and that these microbes were not uniformly distributed in the air. With these four facts now known, Louis Pasteur concluded, that microbes could spread disease among humans. During a cholera epidemic in 1865 Marseilles, France, despite numerous scientific experiments, Louis Pasteur could not isolate the germ that caused the outbreak. It was Robert Koch who first discovered that anthrax microbes produced spores and that the spores had the ability to live for a long time after the host had died, and that the spores could infect other animals. Koch was also able to complete Pasteur's work on cholera. As a result of his infamous four postulates known as Koch's postulates, we have the development of modern microbiology. Koch's postulates state: an organism can be isolated from a host suffering from the disease, the organism can be cultured in the laboratory, the organism causes the same disease when introduced into another host, and the organism can be re-isolated from that host. These postulates still hold true today, that is why Robert Koch is considered the father of the germ theory of disease.

Many atrocities have resulted from the experiments on both animals and humans for a better understanding of the agents and the development of devastating biological weapons. This biological arms race continued until the Biological Weapons Convention, a disarmament treaty considered to be a supplement to the Geneva Protocol, was signed by 150 states and enforced in 1975. The Biological Weapons Convention was intended to prohibit the development, production, and stockpiling of biological or toxin weapons. The lack of formal monitoring of this prohibition has limited the effectiveness of such a convention. The Biological Weapons Convention was suspended in 2001 and negotiations were to resume in 2003. Currently there are 152 states party to the convention. The latest draft resolution proposed by Hungary is available in text format on the United Nations web site.

REFERENCES

Alibek, K. (1999). *Biohazard.* New York: Delta.

Arnon, S. S., Schechter, R., Inglesby, T. V., Henderson, D. A., Bartlett, J. G., et al. (2001). Botulinum toxin as a biological weapon. *Journal of the American Medical Association, 285*, 1059–1070.

Carter, G.B. (2000). *Chemical and biological defence at Porton Down.* London: The Stationary Office.

Carter, G.B. (1992). *Porton Down: 75 years of chemical and biological research.* London: HMSO.

Center for Counterproliferation Research, National Defense University (2001). *Bioterrorism and biocrimes: The illicit use of biological agents since 1900.* Washington, DC: W.S. Carus.

Centers for Disease Control and Prevention. (2000). Use of anthrax vaccine in the United States recommendations of the Advisory Committee in Immunization Practices (ACIP). *MMWR, 49*, (RR-15), 1-20.

Christopher, G.W., Cieslak, T.J., Pavlin, J.A., and Eitzen, E.M. (1997). Biological warfare: A historical perspective. *Journal of the American Medical Association, 278* (5), 412-417.

Georgiades, C., and Fishman, E.K. (2001). Glanders disease of the liver and spleen: CT evaluation. *Journal of Computer Assisted Tomography, 25* (1)

Cieslak, T.J. and Eitzen, E.M. (1999). Clinical and epidemiologic principles of anthrax. *Emerging Infectious Diseases*, 5(4), 552-555.

Covert, N.M. (2000). *The cutting edge: The history of Fort Detrick.* Retrieved on November 11, 2003, from http://www.detrick.army.mil/detrick/cutting_edge/index.cfm

Davis, C.J. (1999). Nuclear blindness: An overview of the biological weapons programs of the former Soviet Union and Iraq. *Emerging Infectious Diseases, 5(4)*, 509-512.

Dembek, Z. (2003). Missed sentinel case of naturally occurring pneumonic tularemia outbreak: Lessons for detection of bioterrorism. *The Journal of the American Board of Family Practice, 16* (4), 339-342.

Dennis, D.T., Inglesby, T.V., Henderson, D.A., Bartlett, J.G., Ascher, M.S., Eitzen, E., et al. (2001). Tularemia as a biological weapon: Medical and public health management. *Journal of the American Medical Association, 285*, (21), 2763-2773.

Diamond, J. (1999). *Guns, germs, and steel: The fate of human societies.* New York: Norton.

Eichner, M. (2003). Analysis of historical data suggests long-lasting effects of smallpox vaccination. *American Journal of Epidemiology, 158*(8), 717-723.

Eitzen, E.M., and Takafuji, E.T. (1997). Historical overview of biological warfare. In R. Zajtchuk (Ed.), *Medical aspects of chemical and biological warfare* (pp. 87-109). Retrieved on September 30, 2005, from http://www.nbc-med.org/SiteContent/HomePage/WhatsNew/MedAspects/Ch-3electrv699.pdf.

Evans, M.E., and Friedlander, A.M. (1997). Tularemia. In R. Zajtchuk (Ed.), *Medical aspects of chemical and biological warfare* (pp. 503-512). Retrieved on September 30, 2005, from http://www.nbc-med.org/SiteContent/HomePage/WhatsNew/MedAspects/Ch-24electrv699.pdf.

Eyler, J. (2003). Smallpox in history: The birth, death and impact of a dreaded disease. *The Journal of Laboratory and Clinical Medicine, 142(4)*, 216-220.

Federation of American Scientists. (n.d.). *Chemical and biological weapons chronology*. Retrieved November 12, 2003, from http://www.fas.org/nuke/control/bwc/chron.htm.

Fenn, E.A. (2000). Biological warfare in eighteenth-century North America: Beyond Jeffrey Amherst. *The Journal of American History, 86(4)*, 1552-1580.

Frischknecht, F. (2003). The history of biological warfare: human experimentation, modern nightmares and lone madmen in the twentieth century [Special issue]. *European Molecular Biology Organization Reports, 4*, S47-S52.

Greenfield, R. (2002). Bacterial pathogens as biological weapons and agents of bioterrorism. *The American Journal of Medical Science, 323(6)*, 299-315.

Guillemin, J. (2005). *Biological weapons: From invention of state-sponsored programs to contemporary bioterrorism.* New York, NY: Columbia University Press.

Henderson, D.A., Inglesby, T.V., Bartlett, J.G., Ascher, M.S., Eitzen, E. et al. (1999). Smallpox as a biological weapon: medical and public health management. *Journal of the American Medical Association, 281*, 2127–37.

Heyman, D.L. (ed.). (2004). *Control of Communicable Diseases Manual, 18th Edition.* Washington, DC: American Public Health Association.

The Holy Bible, revised standard edition. (1952, a). *Exodus 8 and 9*. Philadelphia: Westminster Book Stores.

The Holy Bible, revised standard edition. (1952, b). *I Samuel.* Philadelphia: Westminster Book Stores.

Inglesby, T.V., O'Toole, T., Henderson, D.A., Bartlett, J.G., Ascher, M.S., Eitzen, E., et al. (2002). Anthrax as a biological weapon updated recommendations for management. In Henderson, D.A., Inglesby, T.V., and O'Toole, T. (Eds.), *Bioterrorism guidelines for medical and public health management* (pp. 63-97). Chicago: AMA Press.

Inglesby, T.V., O'Toole, T., Henderson, D.A., Bartlett, J.G., Ascher, M.S., Eitzen, E., et al. (2002). Anthrax as a biological weapon, 2002: Updated recommendations for management. *Journal of the American Medical Association, 287*, 2236-2252.

Inglesby, T.V., Henderson, D.A., Bartlett, J.G., Ascher, M.S., Eitzen, E., Friedlander, A.M., et al. (1999). Anthrax as a biological weapon medical and public health management. *Journal of the American Medical Association, 281, (18)*, 1735-1745.

Jernigan, D.B., Stephens, D.S., Ashford, D.A., Omenaca, C., Topiel, M.S. et al. (2001). Bioterrorism-related inhalational anthrax: The first 10 cases reported in the United States. *Emerging Infectious Diseases, 7(6)*, 933-944.

Jernigan, D.B., Raghunathan, P.L., Bell, B.P., Brechner, R., Bresnitz, E.A. et al. (2002). Investigation of bioterrorism-related anthrax, United States, 2001: Epidemiological findings. *Emerging Infectious Diseases, 8(10)*: 1019-1028.

Kissinger, H.A. (1969). *National Security Memorandum 35*. Retrieved on November 16, 2005, from http://www.gwu.edu/~nsarchiv/NSAEBB/NSAEBB58/RNCBW8.pdf.

Leitenberg, M. (1993). The biological weapons program of the former Soviet Union. *Biologics, 21*, 187-191.

Matsumoto, G. (2004). *Vaccine A: The covert government experiment that's killing our soldier and why GI's are only the first victims*. New York: Basic Books.

McGovern, T.W., and Friedlander, A.M. (1997). Plague. In R. Zajtchuk (Ed.), *Medical aspects of chemical and biological warfare* (pp. 479-501). Retrieved on September 30, 2005, from http://www.nbc-med.org/SiteContent/HomePage/WhatsNew/MedAspects/Ch-23Rprntblscx699.pdf.

Meselson, M., Guillemin, J., Hugh-Jones, M., Langmuir, A., Popova, Sjelokov, A., et al. (1994). The Sverdlovsk anthrax outbreak of 1979. *Science, 266*(5188), 1202-1208.

McClain, D.J. (1997). Smallpox. In R. Zajtchuk (Ed.), *Medical aspects of chemical and biological warfare* (pp. 539-559). Retrieved on September 30, 2005, from http://www.nbc-med.org/SiteContent/HomePage/WhatsNew/MedAspects/Ch-27electrv699.pdf .

McGovern, T.W., and Friedlander, A.M. (1997). Plague. In R. Zajtchuk (Ed.), *Medical aspects of chemical and biological warfare* (pp. 479-501). Retrieved on September 30, 2005, from http://www.nbc-med.org/SiteContent/HomePage/WhatsNew/MedAspects/Ch-23Rprntblscx699.pdf.

Middlebrook, J.L., and Franz, D.R. (1997). Botulinum toxins. In R. Zajtchuk (Ed.), *Medical aspects of chemical and biological warfare* (pp. 643-654). Retrieved on September 30, 2005,from http://www.nbc-med.org/SiteContent/HomePage/WhatsNew/MedAspects/Ch-33Rprntblscx699.pdf.

Migueles, S.A., and Tuazon, C.U. (2001). Anthrax in the new millennium. *Current Treatment Options in Infectious Diseases, (3)*, 247-258.

Miller, J., Engelberg, S., and Broad, W. (2001). *Germs biological weapons and America's secret war*. New York: Simon and Schuster.

Nixon, R.M. (1969). *The President's talking points Congressional leadership meeting* (Tuesday, November 25, 1969). Retrieved on November 16, 2005, from http://www.gwu.edu/%7Ensarchiv/NSAEBB/NSAEBB58/RNCBW10.pdf.

O'Connell, K., Menuey, B., and Foster, D. (2002). Issues in preparedness for biologic terrorism: A perspective for critical care nursing. *American Association of Critical-Care Nurses*, 13(3): 452-469.

Patterson, K.B., and Runge, T. (2002). Smallpox and the Native American. *The American Journal of the Medical Sciences, 323(4)*, 216-222.

Persell, D.J., Arangie, P., Young, C., Stokes, E.N., Payne, W.C., Skorga, P., et al. (2002). Preparing for bioterrorism. *Nursing, 32(2)*: 36-45.

Pittman, P.R, Norris, S.L., Coonan, K.M., and McKee, K.T. (2005). An assessment of health status among medical research volunteers who served in the Project Whitecoat Program at Fort Detrick, Maryland. *Military Medicine, 170 (3)*, 183-187.

Preston, R. (1998, March 9). The bioweaponeers. *The New Yorker*, 52-65.

Real Instituto Elcano. (2004). *Social contact patterns and disease dynamics* (Working Paper (WP) N^0 19/2004). Retrieved November 14, 2005, from http://www.realinstitutoelcano.org/documentos/103/WP-19-2004-I.pdf.

Reilly, C.M., and Deason, D. (2002). Plague: A natural occurring bacterial species can be weaponized. *American Journal of Nursing, 102(11),* 47-50.

Robertson, A.G. (1995). From asps to allegations: Biological warfare in history. *Military Medicine, 160*(8): 369-373.

Tucker, J.B. (2001). *Scourge: The once and future threat of smallpox.* New York: Atlantic Monthly Press.

Varkey, P., Poland, G.A., Cockerill, F.R. 3rd, Smith, T.F., and Hagen, P.T. (2002). Confronting bioterrorism: physicians on the front line. *Mayo Foundation for Medical Education and Research, 77(7)*: 661-672.

Watson, A., and Keir, D. (1994). Information on which to base assessments of risk from environments contaminated with anthrax spores. *Epidemiology and Infection, 113, (3),* 479-790.

Weinstein, R.S., and Alibek, K. (2003). *Biological and Chemical Terrorism a Guide for Healthcare Providers and First Responders* (p. 2). New York: Thieme Medical Publishers, Inc. P. 2.

Wheelis, M. (2002). Biological warfare at the 1346 siege of Caffa. *Emerging Infectious Diseases, 8(9)*, 971-975.

Wheelis, M. (N.D.). A short history of biological warfare and weapons. Retrieved of September 30, 2005, from http://microbiology.ucdavis.edu/faculty/mwheelis/ BW_ history.pdf.

Yom, S. (1997). Plague and AIDS in literature. *Journal of the American Medical Association, 277(5)*: 437-438.

In: Focus on Bioterrorism
Editor: George T. Foster, pp. 67-94
ISBN 1-60021-185-2

Chapter 6

NUCLEAR, BIOLOGICAL, CHEMICAL, AND MISSILE PROLIFERATION SANCTIONS: SELECTED CURRENT LAW

Dianne E. Rennack

ABSTRACT

The use of sanctions specifically to stem weapons proliferation is a relatively new development in U.S. foreign policy. While earlier legislation required the cutoff of foreign aid to countries engaged in specified nuclear proliferation activities and mentioned other sanctions as a possible mechanism for bringing countries into compliance with goals of treaties or international agreements, it was not until 1990 that Congress enacted explicit guidelines for trade sanctions related to missile proliferation. In that year a requirement for the President to impose sanctions against U.S. persons or foreign persons engaging in trade of items or technology listed in the Missile Technology Control Regime Annex (MTCR Annex) was added that year to the Arms Export Control Act and to the Export Administration Act of 1979. Subsequently, Congress legislated economic sanctions against countries that contribute to the proliferation of chemical, biological, and nuclear weapons in a broad array of laws.

This chapter offers a listing and brief description of legal provisions that require or authorize the imposition of some form of economic sanction against countries, companies, or persons who violate U.S. nonproliferation norms. For each provision, information is included on what triggers the imposition of sanctions, their duration, what authority the President has to delay or abstain from imposing sanctions, and what authority the President has to waive the imposition of sanctions.

INTRODUCTION

The use of sanctions specifically to stem weapons proliferation is a relatively new development in U.S. foreign policy.[1] While earlier legislation required the cutoff of foreign aid to countries engaged in specified nuclear proliferation activities and mentioned other sanctions as a possible mechanism for bringing countries into compliance with goals of treaties or international agreements,[2] it was not until 1990 that Congress enacted explicit guidelines for trade sanctions related to missile proliferation. In that year a requirement for the President to impose sanctions against U.S. persons or foreign persons engaging in trade of items or technology listed in the Missile Technology Control Regime Annex (MTCR Annex) was added to the Arms Export Control Act and to the Export Administration Act of 1979. Subsequently, Congress legislated economic sanctions against countries that contribute to the proliferation of chemical, biological, and nuclear weapons in a broad array of laws.

This chapter offers an alphabetic listing and brief description of legal provisions that require or authorize the imposition of some form of economic sanction against countries, companies, or persons who violate U.S. nonproliferation norms.[3] For each provision, information is included on what triggers the imposition of sanctions, their duration, what authority the President has to delay or abstain from imposing sanctions, and what authority the President has to waive the imposition of sanctions.

SELECTED CURRENT LAW: SANCTIONS PROVISIONS

18 U.S.C. (Relating to Criminal Procedure)[4]

18 U.S.C. 229-229F (part I, chapter 11) makes it generally unlawful for a person knowingly "(1) to develop, produce, otherwise acquire, transfer directly or indirectly, receive, stockpile, retain, own, possess, or use, or threaten to use, any chemical weapon; or (2) to assist, induce, in any way, any person to violate paragraph (1), or to attempt or conspire to violate paragraph (1)." The sections establish criminal and civil penalties, and terms of criminal forfeiture.

Sec. 201 of the Chemical Weapons Convention Implementation Act of 1998 (Division I of P.L. 105-277; approved October 21, 1998) enacted these sections to bring the criminal and civil penalties section of United States Code into conformity with the requirements of the Chemical Weapons Convention.

---

1 For a more general discussion on the use of sanctions in foreign policy, see *Economic Sanctions to Achieve U.S. Foreign Policy Goals: Discussion and Guide to Current Law*, Dianne E. Rennack and Robert D. Shuey, CRS Report 97-949; and *Economic Sanctions: Legislation in the 106th Congress*, Dianne E. Rennack, CRS Report RL30384.

2 The International Atomic Energy Act of 1954 and the Nuclear Non-Proliferation Act of 1978 sought to increase international participation in and adherence with the International Atomic Energy Agency and Nuclear Non-Proliferation Treaty, respectively, and, to that end, authorized the President to enter into international discussions, including the imposition of sanctions against those who abrogate or violate these international agreements.

3 The list is arranged alphabetically, with references to U.S. Code and *Legislation on Foreign Relations* where applicable. Legislative history of pertinent amendments is also given, in italics.

4 *Legislation on Foreign Relations Through 1996*, vol. II, page 707.

18 U.S.C. 2332a makes it an offense to use, threaten to use, attempt or conspire to use certain weapons of mass destruction (WMD) against a national of the United States or within the United States. Weapons of mass destruction include a range of destructive devices, defined in 18 U.S.C. 921, and "any biological agent, toxin, or vector." One found to have used a WMD for such use "shall be imprisoned for any term of years or for life, and if death results, shall be punished by death or imprisoned for any term of years of for life."

Sec. 60023(a) of P.L. 103-322 (108 Stat. 1980) added sec. 2332a. The section was substantially reworked by the Antiterrorism and Effective Death Penalty Act of 1996 (P.L. 104-132; approved April 24, 1996). The Chemical Weapons Convention Implementation Act of 1998 (Division I of P.L. 105-277; approved October 21, 1998) exempted chemical weapons from application of this section of 18 U.S.C., and in its place enacted chapter 11B of part I of 18 U.S.C. to establish criminal and civil penalties in conformity with the Chemical Weapons Convention.

Agreement for Nuclear Cooperation between the United States and China, Joint Resolution Approving the Proposed Agreement for Nuclear Cooperation[5]

The agreement requires Presidential certification that China is not violating section 129 of the Atomic Energy Act of 1954, which places restrictions on exports to nations that assist or encourage non-nuclear weapon states to acquire nuclear weapons.

No amendments.

Arms Export Control Act[6]

The Arms Export Control Act (AECA), as amended, authorizes U.S. government military sales, loans, leases, and financing, and licensing of commercial arms sales to other countries. The AECA coordinates such actions with other foreign policy considerations, including nonproliferation, and determines eligibility of recipients for military exports, sales, leases, loans, and financing.

Section 3(f) (Eligibility; 22 U.S.C. 2753(f)) prohibits U.S. military sales or leases to any country that the President determines is in material breach of binding commitments to the United States under international treaties or agreements regarding nonproliferation of nuclear explosive devices and unsafeguarded special nuclear material.

Subsec. (f) was added by sec. 822(a)(1) of the Nuclear Proliferation Prevention Act of 1994 (title VIII of the Foreign Relations Authorization Act, Fiscal Years 1994 and 1995; P.L. 103-236; approved April 30, 1994).

---

5 P.L. 99-183; approved December 16, 1985; 99 Stat. 1174. *Legislation on Foreign Relations Through 1996*, vol. II, p. 1522. On January 12, 1998, the President made the certification required by sec. (b)(1) of this law. The determination also certified that China had met nuclear weapons nonproliferation standards stated in section 902(a)(6)(B)(i) of P.L. 101-246 (22 U.S.C. 2151; often referred to part of the "Tiananmen Square sanctions" in that Act). See Presidential Determination No. 98-10 (63 F.R. 3447; January 23, 1998).

6 P.L. 90-629; approved October 22, 1968; 22 U.S.C. 2751 and following. *Legislation on Foreign Relations Through 1999*, vol. I-A, p. 335.

Section 38 (Control of Arms Exports and Imports; 22 U.S.C. 2778) authorizes the President to control the import and export of defense articles and services, to provide foreign policy guidelines to U.S. importers/exporters, and to promulgate the United States Munitions List constituting what defense articles and services are regulated. Section 38(c) establishes that any person who willfully violates any provision of the section (or of section 39 relating to the reporting of fees, contributions, gifts, and commissions paid by those involved in commercial sales of defense articles or services) may be fined not more than $1 million, imprisoned not more than ten years, or both. Section 38(e) gives the Secretary of State the authority to assess civil penalties and initiate civil actions against violators; any civil penalty for violations under this section is capped at $500,000. Section 38(j) authorizes the President to exempt a foreign country from licensing requirements under the AECA when that country commits to a binding bilateral agreement with the United States to establish export controls on a par with export controls in U.S. law and regulations.

Section 38 was added by sec. 212(a)(1) of the International Security Assistance and Arms Export Control Act of 1976 (P.L. 94-329; approved June 30, 1976). Subsec. (c) was added by the 1976 amendment; the fine and imprisonment terms were amended, however, by sec. 119(a) of the International Security and Development Cooperation Act of 1985 (P.L. 99-83; approved August 8, 1985). Formerly, fine was "not more than $100,000," and period of imprisonment was not more than two years. Subsec. (e) was added by the 1976 amendment. Sec. 119(b) of P.L. 99-83, in 1985, however, added the language that caps civil penalties, and sec. 1303 of the Arms Control, Nonproliferation and Security Assistance Act of 1999 (division B of the Nance/Donovan Foreign Relations Authorization Act, FY 2000-2001; H.R. 3427, enacted by reference in P.L. 106-113), gave civil action authority to the Secretary of State. Previously the section referred to such authority in the Export Administration Act, which resides with the Secretary of Commerce and was capped in that Act at $100,000. Sec. 102(a) of the Security Assistance Act of 2000 (P.L. 106-280; approved October 6, 2000) added subsec. (j).

Section 40 (Transactions With Countries Supporting Acts of International Terrorism; 22 U.S.C. 2780) prohibits exporting or otherwise providing munitions, providing financial assistance to facilitate transfer of munitions, granting eligibility status for such transfers, issuing licenses for such transfers, or otherwise facilitating the acquisition of munitions to a country the government of which "has repeatedly provided support for acts of international terrorism." The section includes in its definition of acts of international terrorism, "all activities that the Secretary [of State] determines willfully aid or abet the international proliferation of nuclear explosive devices to individuals or groups or willfully aid or abet an individual or groups in acquiring unsafeguarded special nuclear material."[7]

The President may rescind the Secretary's determination (sec. 40(f)) by reporting to the Speaker of the House and the Chairman of the Senate Foreign Relations Committee, before issuing the rescission, that the leadership and policies of the country in question have changed, the government is not supporting international terrorism, and the government has issued assurances that it will not support international terrorism in the future. Congress may block the rescission of the terrorist determination by enacting a joint resolution. The President, however, may unilaterally waive any or all of the prohibitions in this section if he

[7] See also sec. 549 of the Foreign Operations, Export Financing, and Related Programs Appropriations Act, 2000.

determines to do so is essential to the national security interests of the United States, and so reports to Congress.

Those found to be in violation of the section face criminal prosecution with penalties of as much as a $1 million fine and imprisonment of not more than ten years. Civil penalties for violations under this section, similar to those in sec. 38, are capped at $500,000; the Secretary of State has the authority to assess civil penalties and initiate civil actions against violators.

Section 40 was added by the Omnibus Diplomatic Security and Antiterrorism Act of 1986 (P.L. 99-399; approved August 27, 1986), and later amended and restated by the Anti-Terrorism and Arms Export Amendments Act of 1989 (P.L. 101-222; approved August 27, 1986). Sec. 822(a)(2)(A) of the Nuclear Proliferation Prevention Act of 1994 (title VIII of the Foreign Relations Authorization Act, Fiscal Years 1994 and 1995; P.L. 103-236; approved April 30, 1994) added a definition of acts of international terrorism that would lead the Secretary of State to make a determination. The same section added definitions "nuclear explosive device" and "unsafeguarded special nuclear material". Sec. 321 of the Foreign Relations Authorization Act, Fiscal Years 1992 and 1993 (P.L. 102-138; approved October 28, 1991) made technical changes to the guidelines for Congress's passage of a joint resolution relating to the section. Sec. 1303 of the Arms Control, Nonproliferation and Security Assistance Act of 1999 (division B of the Nance/Donovan Foreign Relations Authorization Act, FY 2000-2001; H.R. 3427, enacted by reference in P.L. 106-113), gave civil action authority to the Secretary of State. Previously the section referred to such authority in the Export Administration Act, which resides with the Secretary of Commerce and was capped in that Act at $100,000.

Sections 72 and 73 (Denial of the Transfer of Missile Equipment or Technology by U.S. Persons; 22 U.S.C. 2797a; Transfers of Missile Equipment or Technology by Foreign Persons; 2797b), require sanctions against any U.S. citizen or any foreign person whom the President determines to be engaged in exporting, transferring, conspiring to export or transfer, or facilitating an export or transfer of, any equipment or technology identified by the Missile Technology Control Regime (MTCR) that "contributes to the acquisition, design, development, or production of missiles in a country that is not an MTCR adherent..."

Sanctions vary with the type of equipment or technology exported, and are increasingly severe where the type of equipment or technology is more controlled. Worst-case sanctions may be imposed for not less than two years, and include denial of U.S. government contracts, denial of export licenses for items on the U.S. Munitions List, and a prohibition on importation into the United States.

The law allows several exceptions, wherein some or all of the sanctions may not be imposed against foreign persons:

- if an MTCR adherent with jurisdictional authority finds the foreign person innocent of wrongdoing in relation to the transaction;
- if the State Department issues an advisory opinion to the individual stating that a transaction would not result in sanctions;
- if the export, transfer, or trading activity is authorized by the laws of an MTCR adherent and not obtained by misrepresentation or fraud, except when the activity in question is conducted by an entity subordinate to a government of an independent state of the former Soviet Union, and when the President determines that that

government has knowingly transferred missiles or missile technology in a manner inconsistent with MTCR guidelines;

- if the export, transfer, or trade is made to an end-user in a country that is an MTCR adherent;
- in the case of foreign persons fulfilling contracts for defense services or defense articles; then the President will not prohibit importations if
  - the articles or services are considered essential to U.S. national security,
  - the President determines that the provider is a sole supplier and the articles or services are essential to U.S. national security, or
  - the President determines that the articles or services are essential to U.S. national security under defense cooperation agreements or NATO Programs of Cooperation;
- in the case of foreign persons importing products or services into the United States in fulfillment of contracts entered into before the President announces intentions to impose sanctions, then the President will not prohibit importations; or
- in the case of foreign persons providing spare parts, component parts essential to U.S. products or production, routine service and maintenance, essential information and technology.

Sanctions are not imposed, or those imposed may be lifted, against individuals when the President certifies that a foreign government, which is an MTCR adherent, has adequately attended to the violation through some judicial process or enforcement action.

The President may waive the sanction, for either a U.S. citizen or foreign person, if he certifies to Congress that it is essential to the national security of the United States, or that the individual provides a product or service essential to U.S. national security, and that that person is sole provider of the product or service.

Section 1703 of the National Defense Authorization Act for Fiscal Year 1991 (P.L. 101-510; approved November 5, 1990) added sections 71-74. In section 72, sec. 734(a) of the Foreign Relations Authorization Act, Fiscal Years 1994 and 1995 (P.L. 103-236; approved April 30, 1994), added paragraph about "presumption" in guidelines for Presidential determination on transfers of MTCR Annex materials. In sec. 73, sec. 323(a) of the Foreign Relations Authorization Act, Fiscal Years 1992 and 1993 (P.L. 102-138; approved October 28, 1991), added assisting another country in acquiring missiles to the list of sanctionable acts; sec. 1136 of the Arms Control and Nonproliferation Act of 1999 (title XI of the Nance/Donovan Foreign Relations Authorization Act, Fiscal Years 2000 and 2001; H.R. 3427, enacted by reference in P.L. 106-113; approved November 29, 1999) added potential limitation on independent states of the former Soviet Union and the President's certification pertaining to judicial attention by MTCR adherents. Sec. 734(b) of the Foreign Relations Authorization Act, Fiscal Years 1994 and 1995 added the Director of the Arms Control and Disarmament Agency to those with whom the Secretary of State consults when administering the policy. This language, however, was struck out to conform with agency reorganization, particularly that of ACDA being incorporated into the State Department, by sec. 1136 of the Arms Control and Nonproliferation Act of 1999. Sec. 1408 of the National Defense Authorization Act for Fiscal Year 1996 (P.L. 104-106; approved February 10, 1996) made technical changes to reporting requirements relating to issuing a waiver.

Section 73B (Authority Relating to MTCR Adherents; 22 U.S.C. 2797b-2) authorizes the President to impose sanctions against a foreign person, notwithstanding that person's operating in compliance of the laws of an MTCR adherent or that person's exporting to an end-user in a country that is an MTCR adherent, if the country of jurisdiction over that foreign person is a country (1) that has entered into an understanding with the United States after January 1, 2000, (2) for which the United States retains the right to impose sanctions against those in the country's jurisdiction for exporting of controlled items that contribute to the acquisition, design, development, or production of missiles in a country that is not an MTCR adherent.

Sec. 1137 of the Arms Control and Nonproliferation Act of 1999 (title XI of the Nance/Donovan Foreign Relations Authorization Act, Fiscal Years 2000 and 2001; H.R. 3427, enacted by reference in P.L. 106-113; approved November 29, 1999) added sec. 73B, and made supporting amendments in sec. 73 relating to conditions of applicability, and sec. 74, defining "international understanding."

Section 74 (Definitions; 22 U.S.C. 2797c) provides definitions of terms that also affect how the sanctions may be applied. For example, while the MTCR is a policy statement originally announced on April 16, 1987, by the United States, the United Kingdom, Germany, France, Italy, Canada, and Japan, the term "MTCR adherent" in this law is much more broadly defined, to include the countries that participate in the MTCR "or that, pursuant to an international understanding to which the United States is a party, controls MTCR equipment or technology in accordance with the criteria and standards set forth in the MTCR."[8] Within that definition, the term "international understanding" has been further defined to limit its applicability or to make the President's authority to impose sanctions broader. As another example, the term "person" has changed over time. The law formerly included as part of the definition of "person," "countries where it may be impossible to identify a specific governmental entity." This has been amended to refer to "countries with non-market economies (excluding former members of the Warsaw Pact)." The same definition formerly restricted government activity relating to development of aircraft; this now refers specifically to military aircraft.

Sec. 323 of the Foreign Relations Authorization Act, Fiscal Years 1992 and 1993 (P.L. 102-138; approved October 28, 1991) amended the definition of "person" to target China—the "Helms amendment," and narrowed the definition of "person" to include activities of a government affecting the development of, among other things, "military aircraft" (formerly referred to "aircraft"). Sec. 1136(a) of the Arms Control and Nonproliferation Act of 1999 (title XI of the Nance/Donovan Foreign Relations Authorization Act, Fiscal Years 2000 and 2001; H.R. 3427, enacted by reference in P.L. 106-113; approved November 29, 1999) added the definition of "international understanding," a term used in the course of defining "MTCR adherent."

Section 81 ([CBW] Sanctions Against Foreign Persons; 22 U.S.C. 2798) requires imposition of sanctions to deny government procurement, contracts with the U.S. government, and imports from foreign persons who knowingly and materially contribute, through exports from the United States or another country, or through other transactions, to

[8] See also sec. 73A of the AECA (22 U.S.C. 2797b-1), which requires the President to notify Congress when U.S. action results in any country becoming an MTCR adherent. The sections also requires an independent assessment to be submitted to Congress by the Director of Central Intelligence covering the newly designated MTCR adherent and several proliferation issues.

foreign efforts to use, develop, produce, stockpile, or otherwise acquire chemical or biological weapons. Foreign persons are sanctionable if the recipient country has used chemical or biological weapons in violation of international law, has used chemical or biological weapons against its own people, or has made preparations to engage in such violations. Foreign persons are sanctionable if the recipient country has been determined to be a supporter of international terrorism, pursuant to section 6(j) of the Export Administration Act, or if the President has specifically designated the country as restricted under this section.

The President may delay the imposition of sanctions for up to 180 days if he is in consultation with the sanctionable person's government to bring that government to take specific and effective steps to terminate the sanctionable activities. The President may not be required to impose sanctions if the sanctionable person otherwise provides goods needed for U.S. military operations, if the President determines that the sanctionable person is a sole source provider of some good or service, or if the President determines that goods and services provided by the sanctionable person are essential to U.S. national security under defense cooperation agreements. Exceptions are also made for completing outstanding contracts, the purchase of spare or component parts, service and maintenance otherwise not readily available, information and technology essential to U.S. products or production, or medical or other humanitarian items.

The President may terminate the sanctions after 12 months, if he determines and certifies to Congress that the sanctioned person no longer aids or abets any foreign government, project, or entity in its efforts to acquire biological or chemical weapons capability. The President may waive the application of a sanction after a year of its imposition, if he determines it is in U.S. national security interests to do so. Not less than 20 days before a national security waiver is issued, the President must notify Congress, fully explaining the rationale for waiving the sanction.

Sec. 81 was added by sec. 305 of the Chemical and Biological Weapons Control and Warfare Elimination Act of 1991 (title III of P.L. 102-182; approved December 4, 1991.)[9]

Section 101 (Nuclear Enrichment Transfers; 22 U.S.C. 2799aa) (similar to former section 669 of the Foreign Assistance Act of 1961) prohibits foreign economic or miliary assistance to any country that the President determines delivers or receives nuclear enrichment equipment, materials, or technology. The prohibition is not required if the countries involved in the transaction agree to place all materials, equipment, or technology under multilateral safeguard arrangements. The prohibition is not required, furthermore, if the recipient country has an agreement with the International Atomic Energy Agency (IAEA) regarding safeguards.

The President may waive the sanction if he determines, and certifies to the Speaker of the House and the Senate Committee on Foreign Relations, that denying assistance would have a serious adverse effect on vital U.S. interests, and he has been assured that the country in question will not acquire, develop, or assist others in acquiring or developing nuclear weapons. Congress may negate a certification by enacting a joint resolution stating its disapproval.

---

[9] Two versions of the Chemical and Biological Weapons Control and Warfare Elimination Act of 1991 were enacted. Title V of the Foreign Relations Authorization Act, Fiscal Years 1992 and 1992 (P.L. 102-138; approved October 28, 1991) enacted the first. Later in the same session, title III of P.L. 102-182 (a trade act otherwise unrelated to nonproliferation issues) repealed the first version and enacted a new Chemical and Biological Weapons Control and Warfare Elimination Act of 1991. This report refers only to the second enactment—that which currently stands in law.

Sec. 826(a) of the Nuclear Proliferation Prevention Act of 1994 (title VIII of the Foreign Relations Authorization Act, Fiscal Years 1994 and 1995; P.L. 103-236; approved April 30, 1994) added secs. 101 and 102. Similar language, however, previously had been in the Foreign Assistance Act of 1961, as secs. 669 and 670. Sec. 669 was added by sec. 305 of the International Security Assistance and Arms Export Control Act of 1976 (P.L. 94-329; approved June 30, 1976). The section was amended and restated by sec. 12 of the International Security Assistance Act of 1977 (P.L. 95-92; approved August 4, 1977), which also added sec. 670 to the law. Sec. 669 was further amended by secs. 10(b)(4) and 12 of the International Security Assistance Act of 1978 (P.L. 95-384; approved September 26, 1978). Sec. 737(b) of the International Security and Development Cooperation Act of 1981 (P.L. 97-113; approved December 29, 1981) amended and restated both secs. 669 and 670. Sec. 1204 of the International Security and Development Cooperation Act of 1985 (P.L. 99-83; approved August 8, 1985), made further changes to sec. 670 before both sections were repealed in 1994 and similar language was incorporated into the AECA.

Section 102 (Nuclear Reprocessing Transfers, Illegal Exports for Nuclear Explosive Devices, Transfers of Nuclear Explosive Devices, and Nuclear Detonations; 22 U.S.C. 2799aa-1) (similar to former section 670 of the Foreign Assistance Act of 1961) prohibits foreign economic or military assistance to countries that the President determines deliver or receive nuclear reprocessing equipment, material, or technology to or from another country; or any non-nuclear-weapon state that illegally exports, through a person serving as that country's agent, from the United States items that would contribute to nuclear proliferation.

The President may waive the sanction if he determines, and certifies to the Speaker of the House and the Senate Committee on Foreign Relations, that terminating assistance would adversely impact on the United States' nonproliferation objectives, or would jeopardize the common defense and security. Congress may negate a certification by enacting a joint resolution stating its disapproval.

The section further prohibits assistance (except humanitarian or food assistance), defense sales, export licenses for U.S. Munitions List items, other export licenses subject to foreign policy controls (except medicines or medical equipment), and various credits and loans (except Department of Agriculture credits and support to procure food and agriculture commodities) to any country that the President has determined (A) transfers a nuclear explosive device to a non-nuclear-weapon state; (B) is a non-nuclear-weapon state and either (i) receives a nuclear explosive device; or (ii) detonates an nuclear explosive device; (C) transfers to a non-nuclear-weapon state any design information or component that is determined by the President to be important to, and known by the transferring country to be intended by the recipient state for use in, the development or manufacture of any nuclear explosive devices; or (D) is a non-nuclear-weapon state and seeks and receives any design information or component that is determined by the President to be important to, and intended by the recipient state for use in, the development or manufacture of any nuclear explosive device.

In any of these latter four instances, sanctions are mandatory once the President has determined that an event has occurred. If the event has to do with transferring a nuclear explosive device to a non-nuclear-weapon state, or a non-nuclear-weapon state receiving or detonating a nuclear explosive device, the President may delay the imposition of sanctions for 30 days (of congressional continuous session) if he determines that the immediate imposition of sanctions "would be detrimental to the national security of the United States," and so certifies to the Speaker of the House and the Chairman of the Senate Committee on Foreign Relations.

If the President makes such a determination, he may further waive the imposition of sanctions if the Congress, within those 30 days after the first determination, takes up a joint resolution under expedited procedure,[10] that states:

> That the Congress having received on ______ a certification by the President under section 102(b)(4) of the Arms Export Control Act with respect to ______, the Congress hereby authorizes the President to exercise the waiver authority contained in section 102(b)(5) of that Act.

With passage of a joint resolution authorizing him to exercise further waiver authority, the President may waive any sanction which would otherwise be required in instances involving the transferring of a nuclear explosive device to a non-nuclear-weapon state, or a non-nuclear-weapon state receiving or detonating a nuclear explosive device. To exercise this waiver, the President determines and certifies in writing to the Speaker of the House and the Senate Committee on Foreign Relations "that the imposition of such sanction would be seriously prejudicial to the achievement of United State nonproliferation objectives or otherwise jeopardize the common defense and security."

Alternatively, if Congress does not take up a relevant joint resolution within the 30 days, the sanctions enter into effect. Section 102 does not state the means for otherwise suspending or terminating the sanctions.[11]

For legislative history of the origin of and early changes to this section, see discussion following sec. 101, above. Sec. 2(a) of the Agriculture Export Relief Act of 1998 (P.L. 105-194; approved July 14, 1998) broadened the kinds of exchanges that are exempt from the application of sanctions to include medicine, medical equipment, and Department of Agriculture financing.

Atomic Energy Act of 1954[12]

The Atomic Energy Act of 1954 declares U.S. policy for the development, use, and control of atomic energy. The Act authorizes the Nuclear Regulatory Commission to oversee the export of special nuclear materials and nuclear technology in accordance with bilateral and international cooperation agreements negotiated by the Department of State. The Act defines the nature and requirements of those cooperative agreements and the procedure by which Congress reviews them. The Act states export licensing criteria for nuclear materials and sensitive equipment and technology.

---

[10] Sec. 601(b) of the International Security Assistance and Arms Export Control Act of 1976, P.L. 94-329. See *Legislation on Foreign Relations Through 1999*, volume I-A, page 619.

[11] Sanctions under sec. 102 were applied to India and Pakistan after each country tested nuclear explosive devices in May 1998. Congress enacted three laws after the sanctions were imposed to ease their application or authorize the President to waive their application. See the Agriculture Export Relief Act of 1998 (P.L. 105-194; approved July 14, 1998), India-Pakistan Relief Act of 1998 (title IX of P.L. 105-277; approved October 21, 1998), and the Department of Defense Appropriations Act, 2000, title IX (P.L. 106-79; approved October 25, 1999).

[12] P.L. 83-703, approved August 30, 1954, 42 U.S.C. 2011 and following. *Legislation on Foreign Relations Through 1996*, vol. II, p. 1440.

Section 129 (Conduct Resulting in Termination of Nuclear Exports; 42 U.S.C. 2158) prohibits the transfer of nuclear materials, equipment, or sensitive technology from the United States to any non-nuclear-weapon state that the President finds to have detonated a nuclear explosive device, terminated or abrogated safeguards of the International Atomic Energy Agency (IAEA), materially violated an IAEA safeguards agreement, or engaged in manufacture or acquisition of nuclear explosive devices. The section similarly prohibits transfers to any country, or group of countries, that the President finds to have violated a nuclear cooperation agreement with the United States, assisted, encouraged, or induced a non-nuclear-weapon state to engage in certain activities related to nuclear explosive devices, or agreed to transfer reprocessing equipment, materials, or technology to a non-nuclear-weapon state, except under certain conditions.

The President may waive the restriction if he determines that the prohibition would hinder U.S. nonproliferation objectives or jeopardize the common defense and security. Sixty days before a determination is issued, the President is required to forward his reasons for waiving the sanctions to Congress, which may block the waiver by adopting a concurrent resolution. Congress may alternatively counter the Presidential determination with passage of a joint resolution within 45 days of the President's action.

Sec. 307 of the Nuclear Non-Proliferation Act of 1978 (P.L. 95-242; approved March 10, 1978) added sec. 129.

Chemical and Biological Weapons Control and Warfare Elimination Act of 1991[13]

The Chemical and Biological Weapons Control and Warfare Elimination Act of 1991 mandates U.S. sanctions, and encourages international sanctions, against countries that use chemical or biological weapons in violation of international law.

The Chemical and Biological Weapons Control and Warfare Elimination Act of 1991 was enacted as title III of P.L. 102-182 (a) law dealing with trade issues otherwise unrelated to nonproliferation). No amendments have been enacted.[14]

Section 307 (Sanctions Against Use of Chemical or Biological Weapons; 22 U.S.C. 5605) requires the President to terminate foreign assistance (except humanitarian, food, and agricultural assistance) arms sales and licenses, credits, guarantees, and certain exports to a government of a foreign country that he has determined has used or made substantial preparation to use chemical or biological weapons. Within three months, the President must determine and certify to Congress that the government: is no longer using chemical or biological weapons in violation of international law, is no longer using such weapons against its own people, has provided credible assurances that such behavior will not resume, and is willing to cooperate with U.N. or other international observers to verify that biological and

---

[13] P.L. 102-182; approved December 4, 1991; 22 U.S.C. 5601-5606. *Legislation on Foreign Relations Through 1996*, vol. II, p. 1394.

[14] Two versions of the Chemical and Biological Weapons Control and Warfare Elimination Act of 1991 were enacted. Title V of the Foreign Relations Authorization Act, Fiscal Years 1992 and 1992 (P.L. 102-138; approved October 28, 1991) enacted the first. Later in the same session, title III of P.L. 102-182 (a trade act otherwise unrelated to nonproliferation issues) repealed the first version and enacted a new Chemical and Biological Weapons Control and Warfare Elimination Act of 1991. This report refers only to the second enactment—that which currently stands in law.

chemical weapons are not still in use. Without this 3-month determination, sanctions are required affecting multilateral development bank loans, U.S. bank loans or credits, exports, imports, diplomatic relations, and aviation access to and from the United States.

The President may lift the sanctions after a year, with a determination and certification to Congress that the foreign government has met the conditions listed above, and that it is making restitution to those affected by its use of chemical or biological weapons.

The President may waive the imposition of these sanctions if he determines and certifies to Congress and the appropriate committees that such a waiver is essential to U.S. national security interests.

Chemical Weapons Convention Implementation Act of 1998[15]

The Chemical Weapons Convention Implementation Act of 1998 implements the Chemical Weapons Convention, which was originally signed on January 13, 1993, and to which the United States became a party on April 29, 1997.[16] The Convention bans the development, production, stockpiling, and use of chemical weapons, requires the destruction of existing weapons and related materials, establishes an international verification regime, and requires export controls and punitive measures to be leveled for noncompliance.

Section 103 (Civil Liability of the United States; 22 U.S.C. 6713) requires a wide range of sanctions to be imposed, for a period of not less than ten years, on an individual who is a member of, or affiliated with, the Organization for the Prohibition of Chemical Weapons "whose actions or omissions the United States has been held liable for a tort or taking..." or a foreign company or an individual affiliated with that company, "which knowingly assisted, encouraged, or induced, in any way, a foreign person" affiliated with the Organization "to publish, divulge, disclose, or make known in any manner or to any extent not authorized by the Convention any United States confidential business information" including:

- no arms export transactions – sales of items on U.S. Munitions List, transactions under Arms Export Control Act; no licenses for goods or services covered by foreign policy controls under the Export Administration Act of 1979;
- U.S. opposition to support from international financial institutions;
- no U.S. Export-Import Bank transactions;
- prohibition on U.S. private banks engaging with sanctions person;
- assets in United States to be frozen by Presidential action; and
- no rights to land aircraft in the United States (other than in cases of emergency).

The Secretary of State is further required to deny a visa to any individual affiliated with the Organization who divulges any confidential U.S. business if that disclosure results in financial loss or damages.

The section requires the President to impose similar sanctions on any foreign government found by the President to have similarly divulged such information, with the sanctions imposed for not less than five years. Foreign countries are further subject to:

---

[15] Division I of P.L. 105-277; approved October 21, 1998; 112 Stat. 2681-856. See also 18 U.S.C. 229 *et seq.*

[16] See S.Res. 75, 105th Congress, 1st Session.

- no U.S. economic assistance (other than humanitarian assistance), military assistance, foreign military financing, grant military education and training, military credits, guarantees; and no export licensing for commercial satellites.

Sanctions may be suspended if the sanctioned entity fully and completely compensates the U.S. government to cover the liability. The President, alternatively, may waive the sanctions if he determines and notifies Congress that U.S. national security interests are served by such a waiver.

Export Administration Act of 1979[17]

The Export Administration Act of 1979 (EAA) authorizes the executive branch to regulate private sector exports of particular goods and technology to other countries. The EAA coordinates such actions with other foreign policy considerations, including nonproliferation, and determines eligibility of recipients for exports. *Section 5 (National Security Controls; 50 U.S.C. app. 2404)* authorizes the President to curtail or prohibit the export of any goods or services for national security reasons: to comply with other laws regarding a potential recipient country's political status or political stability, to cooperate with international agreements or understandings, or to protect militarily critical technologies. *Section 6 (Foreign Policy Controls; 50 U.S.C. app. 2405)* similarly authorizes the President to curtail or prohibit the export of goods or services for foreign policy reasons. Within section 6, for example, *section 6(j)* establishes the State Department's list of countries found to be supporting acts of international terrorism, a list on which many other restrictions and prohibitions in law are based.[18] *Section 6(k)* restricts exportation of certain crime control equipment. *Section 6(l)* restricts exportation for a list of dual use goods and technology. *Section 6(m)* restricts exportation for a list of goods and technology that would directly and substantially assist a foreign government or group in acquiring the capability to develop, produce, stockpile, or deliver chemical or biological weapons.

Section 11A (Multilateral Export Control Violations; 50 U.S.C. app. 2410a) requires the President to prohibit, for two to five years, the U.S. government from contracting with or procuring goods or services from a foreign person that has violated any country's national security export regulations in accordance with the agreement of the Coordinating Committee for Multilateral Export Controls (COCOM),[19] and that the violation results "in substantial

---

17 P.L. 96-72; approved September 29, 1979; 50 U.S.C. App. 2401. *Legislation on Foreign Relations Through 1996*, vol. III, p. 1022. Authority granted by the Export Administration Act was continued to August 20, 2001, by the Export Administration Modification and Clarification Act of 2000 (P.L. 106-508; approved November 13, 2000).

18 Section 620A of the Foreign Assistance Act of 1961 (22 U.S.C. 2371) and sec. 40 of the Arms Export Control Act (22 U.S.C. 2780) grant similar authority to the Secretary of State to withhold assistance and programs governed by those two laws. See *Legislation on Foreign Relations Through 1999*, vol. I-A, pages 263, 404, respectively. See also sec. 40A of the AECA (22 U.S.C. 2781), which denies defense articles and defense services to a foreign country the President has determined is not fully cooperating with U.S. antiterrorism efforts; *Legislation on Foreign Relations Through 1999*, vol. I-A, page 411.

19 The Coordinating Committee for Multilateral Export Controls (COCOM) agreed to cease to exist on March 31, 1994. Member nations agreed to retain current control lists until a successor organization is established. On December 19, 1995, the United States and 27 other countries, including NATO participants and Russia, agreed to establish a new multilateral export control arrangement. The Wassenaar Arrangement for Export Controls for

enhancement of Soviet and East Bloc capabilities in submarines or antisubmarine warfare, ballistic or antiballistic missiles technology, strategic aircraft, command, control, communications and intelligence, or other critical technologies." The President also is required generally to prohibit importation of products from the sanctioned person. The President may impose sanctions at his discretion if the first but not the second condition exists. In this case, the restrictions may be in place no longer than five years.

Sanctions may not be required for some goods if contracts with the sanctionable person meet U.S. operational military requirements, if the President determines that the sanctionable person is a sole source provider of an essential defense article or service, or if the President determines that such articles or services are essential to U.S. national security under defense coproduction agreements. The President also may not be required to apply sanctions if he determines that a company affiliated with the sanctionable person had no knowledge of the export control violation. After sanctions have been in place for two years, the President may modify terms of the restrictions under certain conditions, and if he notifies Congress.

Sec. 2444 of the Multilateral Export Control Enhancement Amendments Act (title II, subtitle D, part II of the Omnibus Trade and Competitiveness Act of 1988; P.L. 100-418; approved August 23, 1988) added sec. 11A. The section has not been amended.

Section 11B (Missile Proliferation Control Violations; 50 U.S.C. app. 2410b) is similar to sections 72 and 73 of the AECA, but authorizes sanctions against U.S. persons and foreign persons who engage in commercial transactions that violate missile proliferation controls. The section requires sanctions against any U.S. citizen who the President determines to be engaged in exporting, transferring, conspiring to export or transfer, or facilitating an export or transfer of, any equipment or technology identified by the Missile Technology Control Regime Annex. Sanctions vary with the type of equipment or technology exported; worst-case sanctions deny export licenses for goods on the U.S. Commodity List for not less than two years.

The President may waive the imposition of sanctions if he certifies to Congress that the product or service to be restricted is essential to U.S. national security, and that the provider is a sole source provider.

The section further requires sanctions against any foreign person who the President determines to be engaged in exporting, transferring, conspiring to export or transfer, or facilitating an export or transfer of, any MTCR equipment or technology that contributes to the design, development, or production of missiles in a country that is not an MTCR adherent. Sanctions vary with the type of equipment or technology exported; worst-case sanctions deny licenses for transfer to the foreign person items otherwise controlled by the Export Administration Act for not less than two years. The President may also prohibit importation into the United States of products produced by the foreign person.

The law allows several exceptions, wherein some or all of the sanctions may not be imposed against foreign persons. These exceptions are nearly identical to those found in sections 72 and 73 of the AECA. The President may waive the imposition of sanctions for national security reasons, but must notify Congress beforehand. The Presidential authority to restrict importation is conditional in a manner identical to that in section 73 of the AECA.

Conventional Arms and Dual-Use Goods and Technologies ("Wassenaar Arrangement") became operational in 1996. On January 15, 1998, the Bureau of Export Administration of the Department of Commerce issued an interim rule to implement the Wassenaar Arrangement list of dual-use items and revisions to the Commerce Control List required by implementation of the Wassenaar Arrangement. See 63 F.R. 2452.

The definition of "MTCR adherent" in section 11B is also identical to that in section 74 of the AECA. The definition of "person," however, retains its earlier form, applying to all "countries where it may be impossible to identify a specific governmental entity," and not adopting the narrower reference to military aircraft but referring to government activity relating to development of aircraft generally.

Sec. 1702(b) of the National Defense Authorization Act for Fiscal Year 1991 (P.L. 101-510; approved November 5, 1990) added sec. 11B. The section has not been amended.

Section 11C (Chemical and Biological Weapons Proliferation Sanctions; 50 U.S.C. app. 2410c), similar to section 81 of the AECA, authorizes the President to apply procurement and import sanctions against foreign persons that he determines knowingly contribute to the use, development, production, stockpile, or acquisition of chemical or biological weapons by exporting goods or technology from the United States or any other country.

The President may delay the imposition of sanctions for up to 180 days if he is in consultation with the sanctionable person's government to bring that government to take specific and effective steps to terminate the sanctionable activities. The President may not be required to impose or maintain sanctions if the sanctionable person otherwise provides goods needed for U.S. military operations, if the President determines that the sanctionable person is a sole source provider of some good or service, or if the President determines that goods and services provided by the sanctionable person are essential to U.S. national security under defense cooperation agreements. Exceptions are also made for completing outstanding contracts, the purchase of spare or component parts, service and maintenance otherwise not readily available, information and technology essential to U.S. products or production, or medical or other humanitarian items.

The President may terminate the sanctions after 12 months, if he determines and certifies to Congress that the sanctioned person no longer aids or abets any foreign government, project, or entity in its efforts to acquire biological or chemical weapons capability. The President may waive the application of a sanction after a year of its imposition, if he determines it is in U.S. national security interests to do so. Not less than 20 days before a national security waiver is issued, the President must notify Congress, fully explaining the rationale for waiving the sanction.

Sec. 505(a) of the Chemical and Biological Weapons Control and Warfare Elimination Act of 1991 (title III of P.L. 102-182; approved December 4, 1991) added sec. 11C. No amendments have been enacted.

Export-Import Bank Act of 1945[20]

The Export-Import Bank Act of 1945 establishes the Export-Import Bank of the United States and authorizes the Bank to finance and facilitate exports and imports and the exchange of commodities and services between the United States and foreign countries.

Section 2(b)(1)(B) (12 U.S.C. 635(b)(1)(B)) generally states the United States' policy of administering loan programs through the Export-Import Bank. The section provides that the Bank will deny applications for credit for nonfinancial or noncommercial considerations only when the President determines it is in the U.S. national interest to deny credit to advance U.S.

[20] *Legislation on Foreign Relations Through 1996*, vol. III, p. 952.

policies in international terrorism, nuclear proliferation, environmental protection, and human rights.

The Export-Import Bank Act of 1945 was enacted as P.L. 79-173; approved July 31, 1945. Sec. 2(b)(1) has been amended and restated in 1972 (P.L. 92-126) and again in 1974 (P.L. 93-646). The language pertaining to "international terrorism, nuclear proliferation,..." was added by sec. 1904 of the Export-Import Bank Act Amendments of 1978 (title XIX of the Financial Institutions Regulatory and Interest Rate Control Act of 1978; P.L. 95-630; approved November 10, 1978).

Section 2(b)(4) (12 U.S.C. 635(b)(4)) provides that the Secretary of State can determine, and report to Congress and to the Export-Import Bank Directors, if:

- any country has agreed to IAEA nuclear safeguards but has materially violated, abrogated, or terminated such safeguards after October 26, 1977;
- any country has entered into a cooperation agreement with the United States concerning the use of civil nuclear energy, but has violated, abrogated, or terminated any guarantee or other undertaking related to that agreement after October 26, 1977;
- any country has detonated a nuclear explosive device after October 26, 1977, but is a not a nuclear-weapon state;
- any country willfully aids or abets, after June 29, 1994, any non-nuclear-weapon state to acquire a nuclear explosive device or to acquire unsafeguarded special nuclear material; or
- any person knowingly aids or abets, after September 23, 1996, any non-nuclear-weapon state to acquire a nuclear explosive device or to acquire unsafeguarded special nuclear material.

If such a determination is made relating to a person, the Secretary is urged to consult with that person's government to curtail that person's activities. Consultations are allowed 90 days, at the end of which the Secretary will report to Congress as to their progress. After the 90 days, unless the Secretary requests an additional 90 days, or unless the Secretary reports that the violations have ceased, the Ex-Im Bank will not approve any transactions to support U.S. exports to any country, or to or by any person, for which/whom a determination has been made. The imposition of sanctions may also be waived if the President, 45 days before any transaction is approved, certifies that the violations have ceased, and that steps have been taken to ensure the questionable transactions will not resume. The President may also waive the imposition of sanction if he certifies that to impose them would have a serious adverse effect on vital U.S. interests, or if he certifies that objectionable behavior has ceased.

Sec. 2(b)(4) was added by sec. 3(b) of P.L. 95-143; approved October 26, 1977. Sec. 825 of the Nuclear Proliferation Prevention Act of 1994 (title VIII of the Foreign Relations Authorization Act, Fiscal Years 1994 and 1995; P.L. 103-236; approved April 30, 1994) added "(as defined in section 830(4) of the Nuclear Proliferation Prevention Act of 1994), or that any country has willfully aided or abetted any non-nuclear-weapons state (as defined in section 830(5) of that Act) to acquire any such nuclear explosive device or to acquire unsafeguarded special nuclear material (as defined in section 830(8) of that Act)." to define "nuclear explosive device" and to broaden what acts are sanctionable. This is referred to as a "Glenn Amendment." The section was further amended and restated by sec. 1303 of the

National Defense Authorization Act for Fiscal Year 1997 (P.L. 104-201; approved September 23, 1996). Sec. 1303(b) of that Act further required the President to report to Congress within 180 days "his recommendations on ways to make the laws of the United States more effective in controlling and preventing the proliferation of weapons of mass destruction and missiles. The report shall identify all sources of government funds used for such nonproliferation activities."

Section 2(b)(12) (12 U.S.C. 635(b)(12)) requires the President to notify the Export-Import Bank if he determines "that the military or Government of the Russian Federation has transferred or delivered to the People's Republic of China an SS-N-22 missile system and that the transfer or delivery represents a significant and imminent threat to the security of the United States... Upon receipt of the notice and if so directed by the President of the United States, the Board of Directors of the Bank shall not give approval to guarantee, insure, extend credit, or participate in the extension of credit in connection with the purchase of any good or service by the military or Government of the Russian Federation."

Sec. 12 of the Export-Import Bank Reauthorization Act of 1997 (P.L. 105-121; approved November 26, 1997) added paragraph 12.

Foreign Assistance Act of 1961[21]

The Foreign Assistance Act of 1961 (FAA) authorizes U.S. government foreign aid programs including development assistance, economic support funding, numerous multilateral programs, housing and other credit guaranty programs, Overseas Private Investment Corporation, international organizations, debt-for-nature exchanges, international narcotics control, international disaster assistance, development funding for Africa, assistance to states of the former Soviet Union, military assistance, international military education and training, peacekeeping, antiterrorism, and various regional enterprise funds.

Section 307(c) (Withholding of United States Proportionate Share for Certain Programs of International Organizations; 22 U.S.C. 2227) requires that foreign assistance the United States pays in to international organizations and programs not be used for programs in certain countries. The section exempts the International Atomic Energy Agency (IAEA) from this limitation, except for particular projects the IAEA funds in Cuba. U.S. proportionate support to the IAEA, in particular, is not available to any IAEA project relating to the Juragua Nuclear Power Plant near Cienfugeos, Cuba, or the Pedro Pi Nuclear Research Center in Cuba, unless Cuba (I) ratifies the Treaty on Non-Proliferation of Nuclear Weapons or the Treaty of Tlatelelco and is in compliance with terms of the treaty; (II) negotiates full-scope safeguards of the IAEA not later than two years after treaty ratification; and (III) "incorporates internationally accepted nuclear safety safeguards."

Section 307 was added to the Foreign Assistance Act of 1961 by sec. 403 of the International Security and Development Cooperation Act of 1985 (P.L. 99-83; approved

---

[21] P.L. 87-195; approved September 4, 1961; 22 U.S.C. 2151 and following. *Legislation on Foreign Relations Through 1999*, vol. 1-A, p. 15. See also chapter 9 in this Act, relating to "Nonproliferation and Export Control Assistance," added by sec. 301 of the Security Assistance Act of 2000 (P.L. 106-280; approved October 6, 2000), codified at 22 U.S.C. 2349bb *et seq.* This chapter does not impose sanctions; instead it makes assistance available to friendly countries to ultimately "enhance the nonproliferation and export control capabilities...by providing training and equipment to detect, deter, monitor, interdict, and counter proliferation".

August 8, 1985). The countries to which it is applied has changed over time; the countries for which program funding is currently restricted are Burma, Iraq, North Korea, Syria, Libya, Iran, Cuba, and the Palestine Liberation Organization (though application to the PLO has been waived under other legislation in the course of peace negotiations), and communist countries listed elsewhere in the Act. Limitations in subsec. (c) were originally added by sec. 431(a)(2) of the Foreign Relations Authorization Act, 1994 and 1995 (P.L. 103-236; approved April 30, 1994). Language pertaining to nuclear developments in Cuba was added by sec. 2809(a)(1) of the Foreign Relations Authorization Act, 1998 and 1999 (subdivision B of division G of P.L. 105-277; approved October 21, 1998).

Section 498A(b) (Criteria for Assistance to Governments of the Independent States; 22 U.S.C. 2295a(b)) requires that the President not provide assistance to independent states of the former Soviet Union if he determines that the government of that state, among other things, (1) has failed to implement arms control obligations signed by the former Soviet Union, or (2) has knowingly transferred to another country: missiles or missile technology inconsistent with guidelines and parameters of the Missile Technology Control Regime; "any material, equipment, or technology that would contribute significantly to the ability of such country to manufacture any weapon of mass destruction (including nuclear, chemical, and biological weapons) if the President determines that the material, equipment, or technology was to be used by such country in the manufacture of such weapon." The section further prohibits foreign assistance under chapter 11 of the Foreign Assistance Act of 1961 to any country for which a determination has been issued pursuant to sections 101 or 102 of the Arms Export Control Act or sections 306(a)(1) or 307 of the Chemical and Biological Weapons Control and Warfare Elimination Act of 1991.

The President may waive the prohibition – other than that based on other proliferation legislation as cited in the section – on U.S. national security grounds, if he determines that furnishing assistance "will foster respect for internationally recognized human rights and the rule of law or the development of institutions of democratic governance," or to alleviate suffering resulting from a natural or man-made disaster. Assistance may also be provided under the U.S. Information Agency's (USIA) secondary school exchange program notwithstanding a country's ineligibility (except in instances where ineligibility is based on nonproliferation violations). Any waiver requires an immediate report to Congress of any determination or decision.

Section 498A was added by sec. 201 of the FREEDOM Support Act (P.L. 102-511; approved October 24, 1992). See also discussion above, on sec. 73(b)(2) and sec. 73B of the AECA, as amended. Those sections refer to sec. 498A(b)(3)(A) to limit certain transactions with independent states of the former Soviet Union if the transactions involve missiles or missile technology and are conducted in a manner inconsistent with guidelines and parameters of the MTCR.

Section 620(y) (Prohibitions Against Furnishing Assistance; 22 U.S.C. 2370) restricts foreign assistance, or assistance pursuant to any other act, to any country providing nuclear fuel, related assistance, and credits to Cuba. Assistance denied the country in question equals the value of that country's nuclear development assistance, sales, or transfers to Cuba. The requirement to limit assistance is waived if Cuba (A) ratifies the Treaty on Non-Proliferation of Nuclear Weapons or the Treaty of Tlatelelco and is in compliance with terms of the treaty; (B) "has negotiated and is in full compliance with full-scope safeguards of the International Atomic Energy Agency" within two years of the treaty ratification; and (C) "incorporates and is in compliance with internationally accepted nuclear safety safeguards." The section also requires an annual report on the matter to filed with Congress by the Secretary of State.

Added by sec. 2810(a) of the Foreign Relations Authorization Act, Fiscal Years 1998 and 1999 (subdivision B of Division G of P.L. 105-277; approved October 21, 1998).

Section 620A (Prohibition on Assistance to Governments Supporting International Terrorism; 22 U.S.C. 2371) prohibits any foreign assistance, food assistance, Peace Corps funding, and support under the Export-Import Bank Act of 1945 from being made available to countries that the Secretary of State has certified as supporters of international terrorism.[22] The restriction remains in place until such time that the Secretary certifies that there has been a fundamental change in the leadership and policies of the targeted country, the country is no longer supporting international terrorists, and that the targeted government has assured no such support will resume.

The President may waive the prohibition on the basis of U.S. national security, and some assistance may be restored to address humanitarian concerns. A waiver requires notification and justification being provided to Congress 15 days before assistance is given.

Section 620A was added by sec. 303 of the International Security Assistance and Arms Export Control Act of 1976 (P.L. 94-329; approved June 30, 1976). The section has been amended and restated since then by sec. 503(a) of the International Security Assistance and Development Cooperation Act of 1985 (P.L. 99-83; approved August 8, 1985) and sec. 5 of the Anti-Terrorism and Arms Export Amendments Act of 1989 (P.L. 101-222; approved December 12, 1989).

Section 620E (Assistance to Pakistan; 22 U.S.C. 2375), related to U.S. assistance to Pakistan, was enacted in reaction to the threat posed by Soviet occupation of neighboring Afghanistan. *Section 620E(d)* authorizes the President to waive sanctions under section 101 of the AECA to provide assistance to Pakistan, if he determines it is in the U.S. national interest to do so.

Subsection 620E(e) states that no military assistance shall be furnished and no military equipment or technology shall be sold or transferred to Pakistan unless the President certifies to the Speaker of the House and the Chairman of the Senate Foreign Relations Committee that, for the fiscal year in which the assistance, sale or transfer would occur, Pakistan does not possess a nuclear explosive device and that proposed military assistance would significantly reduce the risk that Pakistan will possess a nuclear explosive device. This restriction does not apply to international narcotics control assistance, International Military Education and Training funds, funding for humanitarian and civic assistance projects, peacekeeping or other multilateral operations funds, or antiterrorism assistance.

Enacted as P.L. 87-195; approved September 4, 1961. Sec. 620E was added to the Foreign Assistance Act by sec. 736 of the International Security and Development Cooperation Act of 1981 (P.L. 97-113; approved December 29, 1981). Sec. 620E(d) was amended in 1994 by the Nuclear Proliferation Prevention Act of 1994 (title VIII of the Foreign Relations Authorization Act, Fiscal Years 1994 and 1995; P.L. 103-236; approved April 30, 1994) to reflect the repeal of secs. 669 and 670 and the enactment of secs. 101 and 102 of the Arms Export Control Act. Sec. 620E(e), the "Pressler amendment," was added by

[22] Section 40 of the Arms Export Control Act (22 U.S.C. 2780) and sec. 6(j) of the Export Administration Act (50 U.S.C. app. 2404) grant similar authority to the Secretary of State to withhold assistance and programs governed by those two laws. See also sec. 40A of the AECA (22 U.S.C. 2781), which denies defense articles and defense services to a foreign country the President has determined is not fully cooperating with U.S. antiterrorism efforts.

sec. 902 of the International Security and Development Cooperation Act of 1985 (P.L. 99-83; approved August 8, 1985). Sec. 559(a)(1)(D) of the Foreign Operations, Export Financing, and Related Programs Appropriations Act, 1996 (P.L. 104-107; approved February 12, 1996), amended the section to exclude certain assistance programs from the ban, as noted in the last sentence, above. The same Act made several changes to restrict only "military assistance," formerly the section had referred to assistance generally. The same Act amended the section to authorize the President to: release Pakistan from paying storage costs of items purchased before October 1, 1990, but not delivered (presumably F-16s); release other items serviced in the United States; and continue the applicability of other laws pertaining to ballistic missile sanctions.

After India and Pakistan tested nuclear explosive devices in May 1998, sanctions were imposed in accordance with requirements of sec. 102 of the Arms Export Control Act (see above). Subsequently, Congress enacted three laws to grant the President discretionary authority to waive those sanctions. The third, the Department of Defense Appropriations Act, 2000 (P.L. 106-79; approved October 21, 1999; see title IX), authorized the President to waive section 620E(e) as well.

Section 620G (Prohibition on Assistance to Countries that Aid Terrorist States; 22 U.S.C. 2377) requires the President to withhold all foreign assistance to the government of any country that provides assistance to the government of a country listed as a terrorist state by the Secretary of State pursuant to sec. 620A of this Act (22 U.S.C. 2370).

The President may waive the imposition of the sanction if he determines that furnishing such assistance is important to the U.S. national interest and notifies the appropriate congressional committees of his intent 15 days prior to lifting the ban. His notification shall include the determination, a detailed explanation of the assistance to be provided with its estimated dollar amount, and an explanation of how such assistance furthers U.S. national interests.

Section 620G was added by sec. 325 of the Antiterrorism and Effective Death Penalty Act of 1996 (P.L. 104-132; approved April 24, 1996). See also sec. 40 of the AECA (above).

Section 620H (Prohibition on Assistance to Countries that Provide Lethal Military Equipment to Terrorist States; 22 U.S.C. 2378) requires the President to withhold all foreign assistance to the government of any country that provides lethal military equipment to a country listed by the Secretary of State as a supporter of international terrorism, either on the sec. 6(j) list under the Export Administration Act, or the sec. 620A list pursuant to this Act. The prohibition remains in place until one year after such transfers or transactions cease. The section is not retroactive, but includes all contracts entered into after April 24, 1996 (the date of enactment of the amendment).

The President may waive the imposition of the sanction if he determines that furnishing such assistance is important to the U.S. national interest and notifies the appropriate congressional committees of his intent 15 days prior to lifting the ban. His notification shall include the determination, a detailed explanation of the assistance to be provided with its estimated dollar amount, and an explanation of how such assistance furthers U.S. national interests.

Section 620H was added by sec. 326 of the Antiterrorism and Effective Death Penalty Act of 1996 (P.L. 104-132; approved April 14, 1996).

Foreign Operations, Export Financing, and Related Programs Appropriations Act, 2001[23]

A Foreign Operations Appropriations Act is enacted annually, generally at the start of a fiscal year, to make appropriations for various foreign assistance, military assistance, and international financial institutions programs. Language in the current fiscal year act usually pertains only to that fiscal year. Congress has not enacted a foreign aid authorization bill since 1985, however; as a result, the annual appropriations act increasingly has become a means of enacting authorizing language that carries the force of law beyond the fiscal year.

Title I, Export-Import Bank of the United States, prohibits the use of Export-Import Bank funds in the current fiscal year to be used to make expenditures, contracts, or commitments for the export of nuclear equipment, fuel or technology to any non-nuclear-weapon state, if that state is otherwise eligible to receive economic or military assistance under this Act.

Title II, Assistance for the New Independent States of the Former Soviet Union, appropriates $810 million for assistance to the states of the former Soviet Union but withholds 60 percent of funds obligated for the Government of Russia until the President determines and certifies to the Committees on Appropriations that the Government of Russia has terminated its efforts "to provide Iran with technical expertise, training, technology, or equipment necessary to develop a nuclear reactor, related nuclear research facilities or programs, or ballistic missile capability." The restriction does not apply to assistance for combating infectious diseases or for child survival activities, nor does it apply to nonproliferation and disarmament programs authorized under title V of the FREEDOM Support Act.

Congress has incorporated this language into the foreign assistance appropriations bill for several years. In previous years, the President was authorized to waive the restriction on the basis of vital U.S. national security interests, or if he found that the Government of Russia was taking meaningful steps to limit major supply contracts and to curtail the transfer of technology and technical expertise to certain programs in Iran. The FY2001 Act does not include such a waiver.

Section 549, Prohibition on Assistance to Foreign Governments That Export Lethal Military Equipment to Countries Supporting International Terrorism, prohibits assistance to such foreign governments that provide equipment to any country the government of which the Secretary of State has found, pursuant to sec. 40(d) of the AECA, to be a terrorist government. The President may waive the prohibition if he determines it important to U.S. national interests to do so, but must report to Congress the basis for the determination, the nature and value of resumed assistance, and an explanation of how such assistance furthers U.S. national interests.

Similar language has been included in the annual foreign assistance appropriations measure since FY1994.

[23] H.R. 5526, enacted by reference in P.L. 106-429; approved November 6, 2000. See also sec. 506, Prohibition n Financing Nuclear Goods; sec. 534, Compliance with U.N. Sanctions Against Iraq; and sec. 572, Korean Peninsula Energy Development Organization.

International Emergency Economic Powers Act[24]

Section 203 (Grants of Authorities; 50 U.S.C. 1702) authorizes the President "to deal with an unusual and extraordinary threats with respect to a declared national emergency." After he declares a national emergency exists, pursuant to the authority in the National Emergencies Act, the President may use the authority in this section to investigate, regulate, or prohibit foreign exchange transactions, credit transfers or payments, currency or security transfers, and may take specified actions relating to property in which a foreign country or person has interest. In terms of nonproliferation concerns, it is pursuant to this section that the President has continued the authority of the expired Export Administration Act, prohibited transactions with "those who disrupt the Middle East peace process," issued export controls on encryption items, and established export controls related to weapons of mass destruction.[25]

Enacted as title II of P.L. 95-223; approved December 28, 1977, to update and continue authority carried earlier in the Trading With the Enemy Act (P.L. 65-92; approved October 6, 1917). It has been amended from time to time to update the list of what cannot be restricted, mostly to keep up with changes in technology (for example, the law allows the free flow of informational materials, most recently amended to include CD ROMs).

Iran-Iraq Arms Nonproliferation Act of 1992[26]

Section 1604 (Sanctions Against Certain Persons) requires the President to impose sanctions against any person whom he has determined to be engaged in transferring goods or technology so as to contribute knowingly and materially to the efforts by Iran or Iraq to acquire chemical, biological, nuclear, or destabilizing numbers and types of advanced conventional weapons. *Section 1605 (Sanctions Against Certain Foreign Countries)* similarly addresses activities of foreign governments.

In both cases, mandatory sanctions prohibit, for a period of two years, the U.S. government from entering into procurement agreements with, or issuing licenses for exporting to or for the sanctioned person or country. Where a foreign country is found to be in violation of the law, the President must suspend U.S. assistance; instruct U.S. Executive Directors in the international financial institutions to oppose multilateral development bank assistance; suspend codevelopment and coproduction projects the U.S. government might have with the offending country for one year; suspend, also for one year, most technical exchange agreements involving military and dual-use technology; and prohibit the exportation of U.S. Munitions List items for one year. In the case of foreign countries

---

[24] 50 U.S.C. 1701 and following. *Legislation on Foreign Relations Through* 1996, vol. III, p. 1144.

[25] See Executive Order 12947, Prohibiting Transactions with Terrorists Who Threaten To Disrupt the Middle East Peace Process (January 23, 1995; 60 F.R. 5079; 50 U.S.C. 1701 note); Executive Order 12924, Continuation of Export Control Regulations (August 19, 1994; 59 F.R. 43437; 50 U.S.C. App. 2401 note); Executive Order 12981, Administration of Export Controls (December 5, 1995; 60 F.R. 62981; 50 U.S.C. App. 2401); and Executive Order 13026, Administration of Export Controls on Encryption Products (November 15, 1996; 61 F.R. 58767; 50 U.S.C. App. 2401 note). The President also uses the authority in IEEPA to issue executive orders to implement United Nations Security Council Resolutions. Of the several that are current, Executive Order 12724, Blocking Iraqi Government Property and Prohibiting Transactions with Iraq (August 9, 1990; 55 F.R. 33089; 50 U.S.C. 1701 note) has relevance to issues of proliferation.

[26] 50 U.S.C. 1701 note. *Legislation on Foreign Relations Through 1996*, vol. II, p. 1389.

targeted for sanctions under this Act, the President may, at his discretion, use authority granted him under the International Emergency Economic Powers Act to further prohibit transactions with the country.

The President may waive the mandatory sanctions against persons or foreign country with 15 days notice to congressional committees that exercising such a waiver is essential to U.S. national interests.

Section 1603 (Application to Iran of Certain Iraq Sanctions) makes sanctions in section 586G(a)(1) through (4) of the Iran Sanctions Act of 1990 also fully applicable against Iraq (see below).

Enacted as title XVI of the National Defense Authorization Act for Fiscal Year 1993 (P.L. 102-484; approved October 23, 1992). Sec. 1408(a) of P.L. 104-106 (110 Stat. 494) amended sections 1604 and 1605 to apply not just to conventional weapons but also to chemical, biological, or nuclear weapons.

Iran Nonproliferation Act of 2000[27]

Sections 2 through 5 (Reports; Application; Procedures; Determination; 50 U.S.C. 1701 note) require the President to report to Congress twice a year to identify "every foreign person with respect to whom there is credible information indicating that that person, on or after January 1, 1999, transferred to Iran..." goods, services or technology the export of which (1) is controlled for nonproliferation reasons in accordance with various international agreements, or (2) is not controlled by the country of origin but would subject to controls if shipped from the United States. The President is authorized to apply a range of sanctions against any foreign person included in his report, including denial of procurement contracts with the U.S. government, prohibition on importation into the United States, and denial of foreign assistance – sanctions laid out in Executive Order 12938, as amended.[28] A foreign person named in the President's report may also be denied U.S. government sales of items on the U.S. Munitions List and export licenses for dual-use items.

The decision to impose sanctions is left to the President, but he is required to notify Congress of his reasons to take no action. The President may also take no action if he finds that (1) the person in question did not "knowingly transfer" objectionable items to Iran; (2) the goods, services or technology "did not materially contribute to Iran's efforts to develop nuclear, biological, or chemical weapons, or ballistic or cruise missile systems; (3) the named person falls under the jurisdiction of a government that is an adherent to "one or more relevant nonproliferation regimes" and his actions were consistent with such regime's guidelines; or (4) the government of jurisdiction "has imposed meaningful penalties" on the named person.

---

[27] 114 Stat. 38. See also sec. 708 of the Security Assistance Act of 2000 (P.L. 106-280; 114 Stat. 862; 22 U.S.C. 2797b note; approved October 6, 2000), which requires the President to certify that Russian persons he identifies as "a party to an agreement related to commercial cooperation on MTCR equipment or technology with a United States person" is not also one who transfers goods, services, or technology to Iran, as identified pursuant to sec. 2(a)(1)(B) of this Act.

[28] Executive Order authorizing the Secretaries of Commerce, Treasury, and State to limit or prohibit some transactions to stop the proliferation of weapons of mass destruction. Issued November 14, 1994 (59 F.R. 59099); subsequently amended. See 50 U.S.C. 1701 notes for current text.

Section 6 (Restrictions on Extraordinary Payments in Connection with the International Space Station) prohibits any agency of the U.S. government from making extraordinary payments to the Russian Aviation and Space Agency, or any affiliates, or the Government of the Russian Federation, or any entities of the government, until the President determines and reports to Congress that: (1) it is the Russian government's policy "to oppose the proliferation to Iran of weapons of mass destruction and missile systems capable of delivering such weapons;" (2) the Russian government has demonstrated a commitment to preventing transfers of such goods to Iran; and (3) the Russian Aviation and Space Agency, or its affiliates, has not made such transfers to Iran in the preceding year (other than those allowed by the President's certification for exemptions).

The President may allow extraordinary payments when "such payments are necessary to prevent the imminent loss of life by or grievous injury to individuals aboard the International Space Station." This allowance requires the President to notify to Congress such payments will be allowed, and to report to Congress on details within 30 days of the initial notification. The President may also allow extraordinary payments for specific development programs of the International Space Station provided he notify Congress ahead of payment and that the recipients of that payment are not subject to nonproliferation sanctions.

P.L. 106-168; approved March 14, 2000. It has not been amended.

Iraq Sanctions Act of 1990[29]

This Act reaffirmed the United States' commitment to sanctions leveled by the United Nations after Iraq invaded Kuwait in August 1990. The findings, laid out in *section 586F (Declarations Regarding Iraq's Long-Standing Violations of International Law)*, cite Iraq's violation of international law relating to chemical and biological warfare, Iraq's use of chemical weapons against Iran and its own Kurdish population, efforts to expand its chemical weapons capabilities, evidence of biological weapons development, and its efforts to establish a nuclear arsenal.

Section 586C (Trade Embargo Against Iraq) continues sanctions imposed pursuant to four executive orders issued at the outset of Iraq's invasion of Kuwait. Sanctions include foreign assistance, trade, economic restrictions, and the freezing of Iraqi assets under U.S. jurisdiction. The President may alter or terminate the sanctions issued in his executive orders only with prior 15-day notification to Congress.

Section 586D (Compliance with U.N. Sanctions Against Iraq) prohibits foreign assistance, Overseas Private Investment Corporation (OPIC) funding, and assistance or sales under the AECA to countries found to be not in compliance with United Nations Security Council sanctions against Iraq. The President may waive these sanctions if he determines and certifies to Congress that assistance is in U.S. national interest, that assistance will benefit the targeted country's needy, or such assistance will be in the form of humanitarian assistance for foreign nationals fleeing Iraq and Kuwait.

Section 586G (Sanctions Against Iraq) prohibits the United States from engaging in the following activities relating to Iraq: (1) U.S. foreign military sales under the AECA; (2) commercial arms sales licensing of items on the U.S. Munitions List; (3) exports of control

[29] *Legislation on Foreign Relations Through 1999*, vol. I-B, p. 30.

list goods and technology, as defined by secs. 4(b) and 5(c)(1) of the Export Administration Act; (4) issuance of licenses or other authorizations relating to nuclear equipment, materials, and technology; (5) international financial institutions support; (6) Export-Import Bank funding; (7) Commodity Credit Corporation funding; and (8) foreign assistance other than emergency medical or humanitarian funding.

Pursuant to *section 586H (Waiver Authority)*, the President may waive the application of sec. 586G sanctions if he certifies to Congress that the Government of Iraq has demonstrated improved respect for human rights, does not support international terrorists, and "is not acquiring, developing, or manufacturing (i) ballistic missiles, (ii) chemical, biological, or nuclear weapons, or (iii) components for such weapons; has forsworn the first use of such weapons; and is taking substantial and verifiable steps to destroy or otherwise dispose of any such missiles and weapons its possesses..." The President must further certify that Iraq is meeting its obligations under several international agreements. Finally, the President must certify that it is in the national interest of the United States to make such a waiver and resume any or all of these economic supports. The section also authorizes the President to waive the restrictions in response to a fundamental change in Iraq's leadership, provided the new government makes credible assurances that it meets the above criteria.

Section 586I (Denial of Licenses for Certain Exports to Countries Assisting Iraq's Rocket or Chemical, Biological, or Nuclear Weapons Capability) prohibits the export licensing of supercomputers to any government (or its officials) that the President finds to be assisting Iraq in improving its rocket technology, or chemical, biological, or nuclear weapons capability. While the section includes no waiver authority, it is triggered by the President making a determination and so its implementation rests with the executive branch.

Enacted as secs. 586-586J of the Foreign Operations, Export Financing, and Related Programs Appropriations Act, 1991 (P.L. 101-513; approved November 5, 1990). It has not been amended.

National Emergencies Act[30]

Title II (50 U.S.C. 1621, 1622) authorizes the President to declare, administer, and terminate national emergencies. Such a condition is required for the President to exercise his authority under the International Emergency Economic Powers Act.

P.L. 94-412; approved September 14, 1976. There have been no substantive amendments relevant to proliferation issues.

North Korea Threat Reduction Act of 1999

The North Korea Threat Reduction Act of 1999 prohibits the entering into effect for the United States any international agreement or agreement for cooperation with North Korea that would result in North Korea obtaining nuclear materials. The law also prohibits U.S. issuance of export licenses for, or approval for transfer or retransfer of, nuclear materials, facilities, components, or other goods, services, or technology that would be subject to such an

[30] *Legislation on Foreign Relations Through 1996*, vol. III, p. 1150.

agreement. To make such materials available, the President must determine and report to Congress that North Korea has met certain benchmarks on the safe use of nuclear materials, including: cooperation with the IAEA on inspections, compliance with IAEA safeguard agreements, compliance with terms of the Agreed Framework it reached with the United States, implementation of terms of the Joint Declaration on Denuclearization, no accrual enriched uranium or the means to develop that material, and no efforts to acquire or develop nuclear weapon capability. The President must also determine and certify that it is the U.S. national interest to transfer key nuclear components to North Korea.

Enacted as subtitle B of title VIII of the Admiral James W. Nance and Meg Donovan Foreign Relations Authorization Act, Fiscal Years 2000 and 2001 (H.R.3427, enacted by reference in sec. 1000(a)(7) of P.L. 106-113; 113 Stat. 1501A-472; approved November 29, 1999).

Nuclear Non-Proliferation Act of 1978[31]

The Nuclear Non-Proliferation Act of 1978 states U.S. policy for actively pursuing more effective international controls over the transfer and use of nuclear materials, equipment, and technology for peaceful purposes in order to prevent proliferation. The policy statement includes the establishment of common international sanctions. The Act promotes the establishment of a framework for international cooperation for developing peaceful uses of nuclear energy, authorizes the U.S. government to license exports of nuclear fuel and reactors to countries that adhere to nuclear non-proliferation policies, provides incentives for countries to joint international cooperative efforts in nuclear non-proliferation, and authorizes relevant export controls. The Act requires the Nuclear Regulatory Commission to publish regulations establishing procedures for granting, suspending, revoking or amending nuclear export licenses. The Act also requires the Department of Commerce to issue regulations relating to all export items that could be of significance for nuclear explosive purposes.

Section 304(b) (Export Licensing Procedures; 42 U.S.C. 2155a) requires the Nuclear Regulatory Commission to publish regulations establishing the procedures for granting, suspending, revoking or amending nuclear export licenses. *Section 309 (42 U.S.C. 2139a)* similarly requires the Department of Commerce to issue regulations relating to all export items that could be of significance for nuclear explosive purposes.

Section 402 (Additional Requirements; 42 U.S.C. 2153a) provides that, unless otherwise stated in a cooperation agreement, no source or special nuclear material exported from the United States may be enriched after exportation unless the United States approves the enrichment. The section prohibits the export of nuclear material for the purpose of enrichment or reactor fueling if the recipient country is party to a cooperation agreement with the United States amended or concluded after 1978, unless the agreement specifically allows for such transfers. Finally, the section prohibits export of any major critical component of any uranium enrichment, nuclear fuel reprocessing, or heavy water production facility, unless a cooperation agreement specifically designates these items as exportable.

The Nuclear Non-Proliferation Act of 1978 was enacted as P.L. 95-242; approved March 10, 1978. Secs. 304(b) and 402 have not been amended. Minor changes have been

---

[31] *Legislation on Foreign Relations Through 1996*, vol. II, p. 1417.

incorporated into sec. 309, relating to a requirement of prior consultation and the reorganization of the Department of State.

Nuclear Proliferation Prevention Act of 1994[32]

The Nuclear Proliferation Prevention Act of 1994 was enacted to update current law to reflect growing concerns about nuclear proliferation.

Section 821 (Imposition of Procurement Sanction on Persons Engaging in Export Activities That Contribute to Proliferation; 22 U.S.C. 3201 note) requires U.S. government procurement sanctions against any U.S. person or foreign person if the President determines that that person has materially, and with requisite knowledge, contributed, through export of goods or technology, to efforts to acquire unsafeguarded special nuclear material, or to use, develop, produce, stockpile, or otherwise acquire a nuclear explosive device. Terms of the sanctions are that the U.S. government may not, for 12 months, procure from or enter into procurement contracts with the sanctioned individual. Sanctions may be terminated after 12 months if the President determines and certifies to Congress that the individual has stopped whatever activities that brought on the sanctions, and that the individual will not engage in such activities in the future. Otherwise, to waive the sanctions at the end of 12 months, the President must determine and certify to Congress, 20 days in advance, that continuing the sanctions would have a serious adverse effect on vital U.S. interests.

The President is not required to apply or maintain sanctions if the articles or services provided are essential to U.S. national security; if the provider is a sole source; if the articles or services is essential to national security under defense cooperative agreements; if the articles are essential spare parts, essential component parts, routine servicing or maintenance, or information and technology essential to U.S. production. Sanctions may also not be required if the individual relied on an advisory opinion of the State Department stating that a particular activity was not deemed to be sanctionable.

In the case of a foreign person, the President is required to enter into consultation with the foreign government with primary jurisdiction over that person, and thus may delay the imposition of sanction for up to 90 days. Sanctions may be further averted if the President determines and certifies that the foreign government has taken steps to end the foreign person's activities.

Section 823 (Role of International Financial Institutions; 22 U.S.C. 3201 note) requires the Secretary of the Treasury to instruct U.S. executive directors of international financial institutions to use voice and vote to oppose promotion of the acquisition of unsafeguarded special nuclear material or the development, stockpiling, or use of nuclear explosive devices by any non-nuclear-weapon state.

Section 824 (Prohibition on Assisting Nuclear Proliferation Through the Provision of Financing; 22 U.S.C. 3201 note) prohibits financial institutions and persons involved with financial institutions from assisting nuclear proliferation through the provision of financing. The section requires that when the President determines that a U.S. person or foreign person has engaged in a prohibited activity, he shall impose the following sanctions: (1) ban on dealing in U.S. government debt instruments; (2) ban on serving as a depositary for U.S.

[32] *Legislation on Foreign Relations Through 1996*, vol. II, p. 1356.

government funds; (3) ban on pursuing, directly or indirectly, new commerce in the United States; and (4) ban on conducting business from a new location in the United States.

The President is required to consult with any foreign government that serves as primary jurisdiction for any foreign person sanctioned under this section. Sanctions may be delayed for 90 days while consultation with a foreign government is underway, and may be further averted if the foreign government takes steps to stop the prohibited activity.

Sanctions are in place for not less than 12 months, and are terminated then only if the President determines and certifies to Congress that the person's engagement in prohibited activity has ceased and will not resume. The President may waive the continued use of sanctions when he determines and certifies to Congress that continuing the restrictions would have a serious adverse effect on the safety and soundness of the domestic or international financial system or the domestic or international payments system.

The Nuclear Proliferation Prevention Act of 1994 was enacted as title VIII of the Foreign Relations Authorization Act, Fiscal Years 1994 and 1995 (P.L. 103-236; approved April 30, 1994). Sec. 157(b) of P.L. 104-164 (approved July 21, 1996) made changes to sec. 824, including striking out a requirement that any Presidential determination pursuant to subsec. (c) be reviewed by the courts.

In: Focus on Bioterrorism
Editor: George T. Foster, pp. 95-115

ISBN 1-60021-185-2

Chapter 7

PREVENTING PROLIFERATION OF BIOLOGICAL WEAPONS: U.S. ASSISTANCE TO THE FORMER SOVIET STATES

Michelle Stem Cook and Amy F. Woolf

ABSTRACT

The former Soviet and subsequently Russian biological weapons program possessed capabilities far in excess of any such program known to have existed elsewhere. These capabilities included genetically-altered, antibiotic-resistant pathogens and sophisticated delivery systems. Approximately fifty biological research and production centers (BRPCs) throughout the former Soviet Union devoted either all or part of their work to the program. In the post-Soviet era, former Soviet states drastically reduced and in some cases eliminated funding for these BRPCs. Thousands of BW scientists became unemployed or underemployed, and the facilities, weapons technology, and thousands of strains of pathogens at these BRPCs became vulnerable to theft, sale or misuse.

In the mid-1990s, the United States began engaging BRPCs throughout the former Soviet Union in four kinds of cooperative projects aimed at preventing proliferation of BW capabilities. Collaborative research projects involve former BW scientists in projects with American scientists and seek to deter former BW scientists from selling their expertise to terrorist groups or proliferating states. Several U.S. government agencies are involved in collaborative research projects, most of which are funded through the international science centers. Biosafety enhancement projects are intended to make BRPCs safe places for collaborating scientists to work. In combination, collaborative research and biosafety enhancement projects give U.S. officials routine access to laboratories and facilities that were once used for BW research and production. BioSecurity projects consolidate and restrict access to pathogens. Dismantlement projects target excess infrastructure and BW equipment at BRPC sites for permanent dismantlement. Biosafety, BioSecurity, and Dismantlement projects are funded through and carried out by DOD's Cooperative Threat Reduction Program.

U.S. participants in these projects identify several lessons learned in the past few years. First, it has become clear that the infrastructure of the Soviet/Russian BW complex was more extensive than most analysts realized when the United States initiated its efforts to prevent proliferation of BW capabilities from former Soviet states. Cooperative projects at some BRPCs have helped open doors to other BRPCs, and since 1995, more than forty BRPCs have been involved in cooperative projects with the United States. Second, U.S. participants report that biosafety, biosecurity, and dismantlement projects require complex negotiations, complex engineering work, considerable project management support, and innovative solutions for problems specific to each BRPC. Consequently, they have learned that the United States may need to offer a long-term commitment if it wants to complete the effort. At the same time, the U.S. agencies with BW nonproliferation programs recognize the need to maximize the nonproliferation benefits of U.S. assistance in an environment with limited resources. Finally, U.S. participants have discovered that interpersonal and institutional relationships resulting from these cooperative efforts may play a powerful role in preventing proliferation of BW capabilities from former Soviet states.

ACRONYMS

ARS	Agricultural Research Service of the U.S. Dept. of Agriculture
BRPC	Biological Research and Production Center
BTEP	Biotechnology Engagement Program of the Dept. of Health and Human Services
BW	Biological Weapons
BWC	Biological Weapons Convention
BWPP	Biological Weapons Proliferation Prevention program
CDC	Centers for Disease Control and Prevention
CRDF	Civilian Research and Development Foundation
CTR	Cooperative Threat Reduction
DHHS	Department of Health and Human Services
DOD	Department of Defense
DOE	Department of Energy
DTRA	Defense Threat Reduction Agency
EPA	Environmental Protection Agency
FDA	Food and Drug Administration
FSA	Freedom Support Act
FSU	Former Soviet Union
GAO	General Accounting Office
IPP	Initiatives for Proliferation Prevention Program of the Dept. of Energy
ISTC	International Science and Technology Center
MINATOM	Russia's Ministry of Atomic Affairs, similar to the U.S. Department of Energy
NADR	Non-proliferation, Anti-terrorism, Demining, and Related Activities Account
NIH	National Institutes of Health

NIS	Newly Independent States
NPIR	Non-proliferation Interagency Roundtable
OSD	Office of the Secretary of Defense
SABIT	Special American Business Internship Training
SRCAM	The State Research Center for Applied Microbiology in Obolensk,
STCU	Science and Technology Center in Ukraine
USAMRIID	United States Army Medical Research Institute of Infectious Diseases
USDA	United States Department of Agriculture
USIC	United States Industry Coalition
Vector	The State Research Center for Virology and Biotechnology in Novosibirsk, Russia
WMD	Weapons of Mass Destruction

INTRODUCTION

The events of September and October 2001 undermined long-held assumptions about U.S. vulnerability and the threat of attack with weapons of mass destruction.[1] The attacks on the World Trade Center and Pentagon led many to believe that neither respect for human life nor the will to self-preservation would stop some people from attacking the United States. Then a few letters laced with anthrax caused death, terror, and disruption and exposed the United States' vulnerability to biological attack. Speaking about these events, former Senator Sam Nunn stated, "We lost our sense of invulnerability but as the world is learning, we also lost our sense of complacency."[2]

The United States has been concerned about the potential for proliferation of deadly weapons or materials from the former Soviet Union for more than a decade. These concerns grew in the latter half of 2001. The 107th Congress held several hearings specifically to examine biological weapons programs and the possibility that they might be used against the United States.[3] Testimony often centered on the legacy of the Soviet and subsequent Russian biological weapons (BW) program. The size and scope of that program reportedly dwarfed any BW program known to have existed elsewhere.

This chapter describes the research, development, and production capabilities in the government-sponsored biological weapons complex in the former Soviet Union. It provides an overview of U.S. efforts to prevent proliferation from this complex to other states and sub-state actors. It focuses only on the BW program that was initiated by the Soviet state and sustained by the Russian state. It does not address biological weapons capabilities that may

1 Most analysts consider nuclear, chemical, and biological weapons to be "weapons of mass destruction (WMD)." Some argue that biological weapons are not weapons of mass destruction, but weapons of mass casualties. Others point out that some biological weapons can destroy agricultural and environmental targets on a massive scale.

2 "What changed September 11 was not our vulnerability to terrorism but our understanding of it. The greatest shock was perhaps not even the sheer loss of life, which was staggering, but the evil, hate and fanaticism behind it. To most Americans, the attack was unthinkable. Now our nation knows better. The terrorists' capacity for killing is limited only by the power of their weapons." Sam Nunn, Co-Chairman. Nuclear Threat Initiative. "Toward a New Security Framework." Woodrow Wilson Center, October 3, 2001.

3 Relevant hearings were held in several committees, including the House International Relations Committee, the House Government Reform Committee, the Senate Governmental Affairs Committee—Subcommittee on

exist in former Soviet states outside the state-sponsored biological weapons complex. Furthermore, the report draws only on unclassified sources, so it may not present a complete picture of BW capabilities in former Soviet states or U.S. efforts to prevent their proliferation.

THE SOVIET/RUSSIAN BIOLOGICAL WEAPONS PROGRAM

Background[4]

Reports indicate that Soviet development of biological weapons (BW) dates back to at least 1928 when the Red Army ordered Soviet scientists to find a way to deploy typhus as a weapon. The Soviet BW program expanded in the 1930s; there are strong indications that weapons testing involved some prisoners in Stalin's labor camps. As the program further expanded after the war, the primary known site for BW testing became Vozrozhdeniye Island, which is located in the Aral Sea near Kazakhstan and Uzbekistan.

Negotiations on the 1972 Biological Weapons Convention (BWC) alerted Soviet authorities to Western achievements in weaponizing pathogens.[5] Soviet authorities reportedly doubted that other signatories would abandon their development of biological weapons. At the same time, a prominent Soviet molecular biologist convinced the Soviet leader, Leonid Brezhnev, and other Soviet authorities that developing genetically altered pathogens and more advanced ways to weaponize them would give the Soviet Union a strategic advantage in the Cold War.

In 1973, the Soviet Union created Biopreparat to pursue this new direction in BW development. It was a network of approximately forty "private" biological research and production centers (BRPCs), and reportedly its work remained hidden even from many high-ranking Soviet officials. At its peak, Biopreparat and four military biological institutes reportedly employed approximately 60,000 people. Approximately 9,000 of those were scientists with knowledge that could contribute to the development and production of biological weapons.[6]

Soviet BW Capabilities

According to unclassified reports, the Soviet/Russian BW program developed and produced genetically altered pathogens designed to attack human targets, agricultural targets,

International Security, Proliferation And Federal Services, and the Senate Committee On Appropriations—Subcommittee On Labor, HHS, Education and Related Agencies.

4 For background on the Soviet/Russian Biological Weapons Program see Alibek, Ken, with S. Handelman. *Biohazard: The Chilling True Story of the Largest Covert Biological Weapons Program in the World–Told from the Inside by the Man Who Ran It.* New York: Random House, 1998; Mangold, Tom and Jeff Goldberg. *Plague Wars: The Terrifying Reality of Biological Warfare.* New York: St. Martin's Griffin, 1999; and Miller, Judith, Stephen Engleberg, and William Broad. *Germs: Biological Weapons and America's Secret War.* New York: Simon and Schuster, 2000.

5 For details on the substance of the BWC see *Proliferation Control Regimes: Background and Status.* CRS Report 97-343 F, by Robert D. Shuey, Steve R. Bowman, and Zachary S. Davis.

6 Smithson, Amy. *Toxic Archipelago: Preventing Proliferation from the Former Soviet Chemical and Biological Weapons Complexes.* Washington, DC: The Henry L. Stimson Center, 1999, p. 10.

and environmental targets. The BWC classifies BW agents as bacteria, toxins, viruses, or rickettsaie. The Soviet/Russian program experimented with all these. The program developed and "weaponized" genetically-altered bacteriological agents such as anthrax, plague, tularemia, glanders, and brucellosis that were resistant to heat, cold, and antibiotics.[7] They also "weaponized" several viruses, including:

- Smallpox, which can kill 30-40% of an exposed population during an epidemic;[8]
- Venezuelan Equine Encephalitus, which is unlikely to kill but can incapacitate troops;
- The Marburg virus, which, like the Ebola virus, is a haemorrhagic fever which destroys cells and causes massive internal bleeding.[9]

According to some sources, the Soviet/Russian program also developed "chimera" or "recombinant" weapons, such as a combination of plague bacteria and the myelin toxin which attacks sheaths protecting nerve fibers.[10] Reports indicate that the Soviet/Russian BW program also developed weapons to attack plants and animals. The anti-plant diseases targeted corn, wheat, potatoes, tobacco, barley and other crops.[11] The anti-animal diseases included Foot-and-Mouth Disease and African swine fever.

Reports indicate that the Soviet BW program maintained the capacity to produce and store millions of liters of pathogens. For example, the BW production facility at Berdsk reportedly housed 40 fermenters that could hold 2,560,000 liters of a "weaponizable" pathogen.[12] The Stepnogorsk facility in Kazakhstan had ten twenty-ton fermenters. These could produce approximately 300 tons of anthrax spores in 220 days, reportedly enough to infect the entire population of the United States if effectively dispersed.[13]

The Soviet/Russian BW complex reportedly also succeeded in refining agents to improve the delivery capacity of biological weapons. One ancient way to deliver biological weapons is through the use of infected "vectors," such as bugs, rodents, or cadavers.[14] The Soviet/Russian BW program reportedly developed an infrastructure that could quickly cultivate millions of parasitic insects to carry disease or attack plant life.[15] Biological weapons can also be delivered through contamination of food and water supplies. Most experts agree that the most effective way is to contaminate the air with an aerosolized agent that infects people or animals when they inhale it. The Soviet/Russian program reportedly developed highly-effective techniques to aerosolize agents.[16] Many sources confirm that the

---

7 Mangold and Goldberg, *Plague Wars*, p. 93.

8 *Ibid*, p. 385

9 According to Ken Alibek, one of the Russian weapons scientists died while trying to convert the Marburg virus to a weapon. The Soviets later weaponized the strain of the virus harvested from that scientist's body. See Alibek and Handelman. *Biohazard*, pp. 123-132, 137, 170.

10 *Ibid*, pp. 154-155, 163-164, 166-167

11 Ban, Jonathan. "Agricultural Biological Warfare: An Overview." The Chemical and Biological Arms Control Institute. June, 2000.

12 Mangold and Goldberg, *Plague Wars*, p. 198.

13 See Miller *et al. Germs*, p. 166-167.

14 Zobel, Enrique. "A Matter of Conscience: The real deal about nuclear, biological and chemical attacks," *Business World*. November 14, 2001. p. 4

15 Ban, Jonathan. "Agricultural Biological Warfare: An Overview." June 2000.

16 Alibek, Ken. Address at a conference entitled "Globalization and Infectious Diseases: Institutions, Policies, and the Threat of Bioterrorism." Held at the Johns Hopkins University School of Advanced International Studies (SAIS) in association with Novartis Corporation, Washington, DC, November, 16, 2001.

Soviet/Russian BW program refined delivery systems for biological agents by conducting tests on primates at Vozrozhdeniye Island.[17] The Soviet Air Force reportedly equipped some aircraft specifically to carry out biological attacks. One type of airplane had the capacity to spray biological agents. Another was designed to carry cluster bombs, with cantaloupe-sized bomblets that could spin and spread disease as they fell from the sky.

Revelations about the Soviet/Russian BW Program

In 1979, at least sixty-eight people died when anthrax spores were released from a Biopreparat production facility in Sverdlovsk, Russia.[18] Although suspicions were raised from 1979 onward, the United States did not verify that the deaths had been connected with the Soviet biological weapons program until 1992.[19] Some information became available in 1989, when Vladimir Pasechnik, the former director of a Biopreparat facility, defected to Great Britain. He revealed that the magnitude and lethality of the Soviet/Russian BW program far exceeded the assessments by U.S. and British intelligence sources.[20]

Early in 1992, Russia's President Boris Yeltsin publicly acknowledged the existence of an active Russian BW program. He vowed to end the program, but his efforts were reportedly thwarted by some in his military establishment, who continued the program. In September of 1992, these military officials represented Russia in negotiations with the United States and Great Britain that sought to establish an inspection process to confirm the absence of BW production. Russia initially consented to U.S. and British inspections at Russian BW facilities, but then refused to sign the "Trilateral Agreement" unless the United States and Great Britain accepted reciprocal inspections at private pharmaceutical companies in their countries. Some in Russia claimed that the United States and Great Britain were violating the BWC. After inspecting U.S. facilities, Russian inspectors claimed they found evidence of an American BW program. Yet Russia never gave U.S. and British inspectors access to military BRPC's in Russia.[21]

In 1992, Kanatjan Alibekov, who has changed his name to Ken Alibek, defected to the United States. He had been second in command in the Soviet/Russian BW program. In his debriefings with U.S. officials, Alibek confirmed and expanded on Pasechnik's revelations. Alibek has frequently testified before Congress about the Soviet/Russian BW program and about U.S. efforts to mount a defense against biological attacks.[22]

---

[17] Mangold and Goldberg, *Plague Wars*, pp. 94-95.

[18] Alibek and Handelman. *Biohazard*. pp. 70-86.

[19] Miller *et al. Germs* pp. 79-82, 93-94, 134-135, 143-144, 175, 178, 221. See, also, *Anthrax at Sverdlovsk: U.S. Intelligence on the Deadliest Modern Outbreak* National. Security Archive Electronic Briefing Book No. 61 Edited by Robert A. Wampler and Thomas S. Blanton November 15, 2001.

[20] Obituary of Vladimir Pasechnik. *The Times Newspapers*, November 21, 2001.

[21] Mangold and Goldberg, *Plague Wars*, 170-176.

[22] Although Alibek is frequently cited as a source for information about Russia's BW program, some U.S. experts have challenged his assertions. They have focused, specifically on his claims about Soviet/Russian success with combining Ebola and smallpox, and his assessment of equipment that was discovered in a BRPC in Kazakhstan. Miller *et al. Germs*, p. 220, 226, 292-293.

The Threat Posed by the Soviet/Russian BW Program

Many observers believe that the legacy of the Soviet/Russian BW program might pose a significant threat both to the Russian people and to the security of the United States. There are several sources for these concerns.[23]

First, experts fear that the contamination at Vozrozhdeniye Island could pose a threat to surrounding areas. Experts believe that the ground on the island has absorbed biological agents scattered during weapons tests. Furthermore, the Aral Sea is shrinking, and some fear that rodents might soon be able to travel between the island and the mainland, possibly spreading disease to a population already sickened by environmental degradation and pollution.[24] Natural environments surrounding many other BRPCs may also be degraded due to widespread environmental pollution and, possibly, the leakage of some biological weapons agents into the soil.

Second, reports indicate that many BRPCs in former Soviet states are in bad physical condition. These BRPCs have not been able to maintain advanced biosafety containment laboratories in the post-Soviet economic environment. Experts fear that accidental release of pathogens could occur at many of these sites.[25] Nevertheless, some experts believe that, despite the degraded condition of many BRPCs, Russia could use the remaining expertise and infrastructure to reactivate an offensive biological weapons program.

Third, many experts believe that biological weapons capabilities in former Soviet states could be vulnerable to theft or sale.[26] There are reports that the mafia and warring ethnic factions within Russia have tried to obtain biological weapons capabilities.[27] Personnel at BRPCs in former Soviet states are generally poorly paid, which, some believe, could motivate them to steal and sell dangerous pathogens, weapons technologies, or instructions related to BW development and production.[28] Seed cultures of pathogens might also be smuggled out of the BRPCs.

The United States does not have a complete knowledge of the biological weapons capabilities in the Soviet/Russian BW complex, and therefore, may not have vaccines or antibiotics that could provide a defense against infection. U.S. biodefense efforts have increased since September 11, 2001, but many public health officials stress that U.S. preparedness for a biological attack is still inadequate. In addition, according to many experts, biological weapons have several characteristics that could make them attractive to terrorist groups and hostile nation-states – they can produce mass casualties and incite panic;[29] it could be difficult to trace the perpetrator of a biological attack;[30] and they could provide an

---

[23] Tucker, Jonathan. "Bioweapons from Russia, Stemming the Flow." *Issues in Science and Technology Online.* Spring 1999.

[24] Miller *et al. Germs*, p. 180-182

[25] U.S. General Accounting Office. "Effort to Reduce Former Soviet Threat Offers Benefits, Poses New Threat." GAO/NSIAD-00-138. April 2000.

[26] *Ibid.*

[27] Alibek and Handelman, *Biohazard*, pp. 176-177; See also, Miller *et al. Germs*, p. 211

[28] Smithson, *Toxic Archipelago.*

[29] "Chemical and biological weapons can produce panic more disastrous than the agents themselves, and are much easier to develop and conceal." Barclay, Glen. "Bioweapons are the most deadly threat." *Courier Mail.* Nationwide News Pty Limited. November 8, 2001, p. 17.

[30] Alibek and Handelman, *Biohazard* p. 176.

"asymmetric means" of challenging "America's overwhelming conventional and nuclear war-fighting strength."[31]

Some sources report that representatives of Iran and Al-Quaeda[32] have attempted to recruit former Soviet scientists with BW expertise. According to George Tenet, the Director of Central Intelligence, these scientists possess expertise that could confer "the advantage of technological surprise" on enemies of the United States.[33] Furthermore, hostile nations and terrorist groups might offer salaries significantly higher than the scientists' current, often very low, incomes.[34]

According to some observers, this problem might be overstated. Although exact statistics are not available, the vast majority of Soviet weapons scientists apparently either remained in former Soviet states or migrated to the United States and Britain. The desire of remaining Soviet weapons scientists to stay rooted in their home communities near relatives and friends could make it difficult for proliferating groups to lure them elsewhere. However, a Carnegie Endowment report on nuclear and missile scientists, released in April 2001, indicated that although many former weapons scientists once hoped the economic crisis would pass, now some of them may migrate because they fear their situation will not improve. Some scientists also apparently feel that it would be better to be paid to work for a proliferating state or sub-state group than to become involved in organized crime.[35] In addition, BW scientists could still serve the interest of proliferators without traveling or migrating if they provided consultation, shared weapons secrets over the internet, or stole a tiny seed culture of a genetically-altered pathogen and passed it into the hands of proliferators.[36]

THE U.S. RESPONSE

Policy Objectives for Nonproliferation Efforts

Since 1995, the United States has gained access to many BRPCs in former Soviet states and has sought to prevent proliferation of BW capabilities. These efforts are directed toward several policy objectives, including:

- Increasing transparency at BRPCs that once participated in the Soviet/Russian BW program;
- Securing or destroying pathogens and weapons technology so that they are not sold, stolen, accidentally deployed or leaked, or used to reactivate a biological weapons program in Russia and/or other former Soviet states;

---

[31] U.S. Senate, Committee on Governmental Affairs, Subcommittee on International Security, Proliferation, and Federal Services. Statement of Dr. Anna Johnson-Winegar, Deputy Assistant to the Secretary of Defense for Chemical and Biological Defense. October 17, 2001.

[32] Miller *et al. Germs*, pp. 205-207, 209-212, 228-229, 280. See, also, "Evidence Indicates Al Qaeda Had Russian Help Developing Anthrax." New York. *PRNewswire*. December 9, 2001.

[33] Miller *et al. Germs*, p. 287.

[34] *Ibid*, pp. 205-206.

[35] Tikhonov, Valentin. *Russia's Nuclear and Missile Complex: the Human Factor in Nonproliferation*. A Report by the Non-proliferation Project of the Carnegie Endowment for International Peace. 2001

[36] Smithson, Amy. *Toxic Archipelago*.

- Redirecting biological weapons scientists towards peaceful pursuits and discouraging them from interacting with terrorist groups or proliferating states;
- Strengthening the United States' preparedness for biological attack.

Two secondary goals are strengthening the scientific research and development capability of former Soviet states, and reinforcing "the transition to market-based economies responsive to civil needs."[37] The United States funds and administers several programs that are designed to meet these objectives; these are summarized on the Table that appears at the end of this chapter.

The State Department

The Science Centers

In 1992, the United States, Japan, the European Union, and Russia established the International Science and Technology Center (ISTC) "to develop, approve, finance, and monitor science and technology projects for peaceful purposes" in former Soviet states. The Russian parliament initially resisted the establishment of the ISTC, but the Center began making grants to nuclear, chemical, and biological weapons scientists in 1994.[38] A similar organization called the Science and Technology Center in Ukraine (STCU) was founded in July 1995. It distributes grants to former weapons scientists in Ukraine, Uzbekistan, and Georgia. The STCU and the ISTC have tax-exempt status and their employees have diplomatic status. Scientists who receive these grants may continue to pursue their regular jobs, but these grants seek to provide them with enough income to reduce their incentive to sell their knowledge to other nations.

In FY2002, the Administration requested and Congress appropriated $37 million to the ISTC and STCU. When making grants to former Soviet scientists, the ISTC and STCU seek to prevent the misappropriation of funds. According to the GAO, "Since 1994, the International Science and Technology Center has directly deposited grant payments into project participants' individual bank accounts, which prevents the institutes from diverting funds for unauthorized purposes."[39] In addition, both centers have the right, the responsibility, and the power to audit the grants they make.[40]

The Science Centers provide a framework for various agencies in the United States government to engage in collaborative research with former weapons scientists. The Department of State serves as the lead agency. As is noted below, the Departments of Health and Human Services, Agriculture, Energy, and Defense all fund projects through the ISTC and STCU to engage Russia's biological weapons scientists. The Environmental Protection Agency is also developing some collaborative research proposals.

Private companies, universities and other private organizations help develop Science Center projects into sustainable, commercially competitive enterprises. The ISTC Partner

[37] Department of Defense. Cooperative Threat Reduction Program. Biological Weapons Proliferation Prevention.
[38] Smithson, Amy, *Toxic Archipelago* pp. 22-46. For a detailed account of the establishment and functioning of the ISTC, see Schweitzer, Glenn E. *Moscow DMZ: the Story of the International Effort to Convert Russian Weapons Science to Peaceful Purposes*. Armonk, N.Y.: M. E. Sharpe. 1996.
[39] GAO/NSIAD-00-138.
[40] Amy Smithson. *Toxic Archipelago*, pp. 32-33.

Program facilitates collaboration between private organizations and BRPCs in former Soviet states, and provides private organizations with the same advantages, such as tax-exempt status and auditing capabilities, that it provides U.S. government agencies. According to the State Department, this program has contributed over $20 million annually to the ISTC and STCU.[41]

Science Center projects also include scientists in former Soviet states who did not participate in the Soviet/Russian BW program. According to the U.S. participants, incorporating non-weapons scientists promotes the goal of integrating former weapons scientists into the mainstream scientific community.[42]

Redirection of Biotechnical Scientists Program

Congress has appropriated funding for the Redirection of Biotechnical Scientists Program through the State Department, under the Freedom Support Act (P.L. 102-511). The Administration has requested $20 million for this program for FY2003. In prior years, funding for the program was in the State Department's Newly Independent States (NIS) Account. Beginning in FY2003, funding will be in the Non-proliferation, Anti-terrorism, Demining, and Related Activities (NADR) Account. The State Department allocates this funding among programs in the Departments of Agriculture, Health and Human Services, and the Environmental Protection Agency.

Department of Agriculture

In 1998, the Agricultural Research Service (ARS) of the United States Department of Agriculture (USDA) became involved in the Redirection of Biotechnical Scientists Program by launching the ARS-Former Soviet Union Scientific Cooperation Program. ARS scientists collaborate with former BW scientists who have expertise in animal and plant diseases. This program is intended to:

- Establish collaborative, mutually beneficial research;
- Maintain substantial contact between FSU and ARS scientists through reciprocal visits;
- Optimize collaboration at the scientist level and share success between both partners.[43]

As of December 2001, ARS had nine on-going projects in Russia and four on-going projects in Kazakhstan. The program's U.S. participants were also developing new projects with institutes in these two nations. Twenty-three proposals were under review involving four former BW institutes in Uzbekistan. The ARS-Former Soviet Union Scientific Cooperation Program received $550,000 for FY1998, $2 million for FY1999, $6.98 million for FY2000, $6 million for FY2001 and expects $5 million in FY2002.

---

[41] Statement of Deputy Assistant Secretary of State Vann Van Diepen before the Senate, Committee on Governmental Affairs, Subcommittee on International Security, Proliferation, Federal Services. Hearing on the Nonproliferation Assistance Coordination Act. November 19, 2001.

[42] HHS briefing for the National Research Council's Committee on Assessing Research Proposals and Projects of Russian Biology Institutes.

[43] USDA briefing for the National Research Council's Committee on Assessing Research Proposals and Projects of Russian Biology Institutes.

Department of Health and Human Services

At the request of the State Department and Department of Defense, the Department of Health and Human Services (HHS) established the Biotechnology Engagement Program (BTEP) in March 1999. The program focuses on biodefense research and high-priority public health problems in former Soviet states. Experts argue that this program can benefit the United States because some Soviet/Russian innovations, such as aerosolized vaccines for mass immunization,[44] might strengthen U.S. preparedness for a biological attack.[45] American scientists who participate in BTEP projects come from the U.S. Army Medical Research Institute of Infectious Diseases (USAMRIID), U.S. academic institutions, and HHS agencies, including the Centers for Disease Control and Prevention (CDC), the National Institutes of Health (NIH), and the Food and Drug Administration (FDA).[46] As of November 2001, BTEP had completed two projects, had nine ongoing projects, sixteen projects in the final stages of authorization, and an additional twenty-nine proposals under review. Plans for FY2002 include development and implementation of the Applied Epidemiology Training Program and developing a "grantsmanship" training program for researchers in former Soviet states.[47] The Biotechnology Engagement Program (BTEP) received $4.8 million for FY1999, with a 50% restriction on funding to Russia, $11 million in FY2000, $10 million in FY2001 and $9 million in FY2002. The budget includes a request for $10 million for FY2003.

Environmental Protection Agency

The Environmental Protection Agency is also becoming involved in collaborative research with former BW scientists; these projects are exploring the environmental effects of BW. It received $1.02 million for its involvement in the Redirection of Biotechnical Scientists Program in FY2001. It has allocated approximately $1 million to specific projects, but has not yet notified Congress of these obligations.

Department of Energy

The Department of Energy's (DOE) Initiatives for Proliferation Prevention (IPP) Program unites scientists from the ten U.S. National Laboratories with industry partners and former nuclear, chemical, and biological weapons scientists in former Soviet states.[48] Some IPP programs are funded through the ISTC, others are not.[49] IPP seeks to redirect underemployed and unemployed former Soviet weapons scientists to commercial ventures by

---

44 Miller *et al. Germs*, p. 180.

45 McMurray, Jeffrey. "Nunn: USSR Weapon-Makers May Help." *Associated Press*. October 25, 2001.

46 HHS briefing for the National Research Council's Committee on Assessing Research Proposals and Projects of Russian Biology Institutes.

47 U.S. Department of Health and Human Services. BTEP Annual Report, FY2001.

48 U.S. Department of Energy. Budget Request for FY 2000; Hearing before the U.S. House of Representatives Committee on National Security Subcommittee on Military Procurement, March 4, 1999. See also, Chase, Marilyn. "Turning Swords into Plowshares," *Wall Street Journal*. P. B1. November 20, 2001.

49 When IPP projects are funded outside the framework of the ISTC or STCU, a tax-exempt organization serves as a contractor to DOE and deposits grant money in the accounts of individual scientists in the same way the Science Centers do.

involving them in market-driven collaboration with U.S. labs and industry partners. The involvement of industry partners makes IPP collaborative research projects different from those sponsored by HHS, the EPA, and the USDA. In FY2002, Congress appropriated $54 million ($39 million in the Energy and Water Appropriations Bill and $15 million in the Emergency Supplemental Appropriations Bill), for the Russian Transition Initiative, a program that combines IPP with DOE's Nuclear Cities Initiative. In recent years, roughly 20% of IPP funding has gone towards the BW engagement effort,[50] which is relatively new compared to the IPP nuclear effort. As of November 2000 according to DOE, the IPP Program had "engaged 20 biological institutes and almost 600 scientists, approved more than 55 projects, and allocated over $12 million for collaboration with former biological weapons facilities."[51] DOE hopes that several IPP biological projects might bring their results to market in the near future.

Department of Defense

The Department of Defense (DOD) Defense Cooperative Threat Reduction Program (CTR) includes the Biological Weapons Proliferation Prevention Program (BWPP). The Administration requested, and Congress appropriated, $17 million for this effort in FY2002. It has three major components: biosecurity and biosafety enhancements, collaborative research, and facilities and equipment dismantlement.

Biosafety and Biosecurity Enhancement Projects seek to ensure the secure and safe storage and handling of biological pathogens at biological research centers.[52] They seek to counter both potential outside and inside actors who might steal or sell BW capabilities. These projects also seek to establish a dialogue between U.S. and former Soviet scientists to allow for the identification of facilities and equipment that could be dismantled. There are currently at least six DOD biosafety and biosecurity projects, with six more in the planning stages. The ongoing projects will institute integrated systems to consolidate, store, secure, and account for pathogens. The projects will also develop and implement programs that restrict personnel access and hold personnel accountable for their activities inside the facilities. DTRA explicitly avoids "state of the art" systems and instead institutes standard, industry-proven systems.[53] DOD funds and implements the BioSafety and BioSecurity Enhancement projects through the ISTC and STCU. Costs for completed BioSafety and BioSecurity Enhancement projects are expected to range from $5 million for a small BRPC to $15 million for a large BRPC.[54]

The Collaborative Research Program is designed to "prevent the proliferation of BW technologies, increase transparency, and enhance US force protection capabilities through research projects with former BW scientists at the BRPCs."[55] Through 2000, DOD had funded projects that "employed more than 350 former biological weapons scientists from

---

50 Interview with IPP official.

51 "Energy Department's Idaho Lab Teams with Russia to Establish Ecological Biotrade Center." DOE Press Release No. R-00-287. November 14, 2000.

52 Defense Threat Reduction Agency, Cooperative Threat Reduction Projects in Russia.

53 Defense Threat Reduction Agency. Biological Weapons Proliferation Prevention project plans.

54 Ibid.

55 Ibid.

seven institutes in bio-defense and public health projects of interest to the United States."[56] The BWPP collaborative research program has significantly expanded since then. On average, the total cost of a BWPP Collaborative Research Project is approximately $700,000, inclusive of project management costs, logistical costs, compensation of scientific collaborators and all other costs.

The BWPP Dismantlement Projects are designed to complicate reconstitution of a BW program and prevent proliferation of BW technology by eliminating infrastructure and equipment at biological research and production centers. The first major dismantlement project occurred at Stepnogorsk, the massive BRPC in Kazakhstan. The program is dismantling the entire facility. The project is currently in its final phase, with the total cost projected to be $10 million. Future dismantlement projects at other BRPCs will eliminate BW equipment, enable BRPCs to lower operational costs by consolidating their operations in fewer structures, and prepare the BRPCs for collaboration with U.S. industry partners. The United States has also agreed to try to eliminate the contamination remaining on Vozrozhdeniye Island, the Soviet BW testing site in the Aral Sea.

Department of Commerce

The Department of Commerce indirectly supports interagency BW nonproliferation efforts through its Special American Business Internship Training (SABIT) program. The SABIT program was instituted in 1990 to support economic restructuring in former Soviet states, and it funds short-term training at U.S. companies for specialists from former Soviet states. SABIT also offers courses in Russia that provide participating groups with an assessment of their prospects for product commercialization and market success. The Department of Commerce also provides BISNIS, the Business Information Service for the Newly Independent States.

Non-Governmental Organizations (NGOs)

The United States Industry Coalition (USIC), a non-profit association of U.S. companies and universities, works to facilitate industry involvement in U.S. nonproliferation efforts in former Soviet states by serving as an advocate for their interests.[57] For example, USIC strives to ensure the Intellectual Property Rights of its member organizations. Universities and companies involved in IPP programs are required to be members of the USIC.

The Civilian Research and Development Foundation for the Independent States of the Former Soviet Union (CRDF), is a non-profit organization created by the United States in 1995 to facilitate mutually-beneficial scientific and technical collaboration between the United States and former Soviet states. The CRDF is funded through government contracts and several private foundations to provide services such as the administration of grants,

[56] Ibid.

[57] USIC's revenues from government contracts were $737,987 in FY2000 and $925,077 in FY2001. See USIC Annual Report 2000-2001. See also, U.S. Department of Energy. "Department of Energy Announces Consumer and Environmental Projects With Former Soviet Biological Weapons Scientists and U.S. Industry Partners." July 24, 2000.

organization of meetings, and merit-based technical review of proposals. Through its Partner Program, the CRDF matches U.S. companies with scientists and engineers in former Soviet states who have skills the companies need. The CRDF operates a Collaborative Research Program entirely devoted to nonproliferation of BW capabilities from former Soviet states; this program primarily supports the Collaborative Research component of DOD's Biological Weapons Proliferation Program.

The National Academies have actively supported U.S. efforts to prevent proliferation of BW capabilities by providing advice on matters related to science and technology. For example, the Committee on Assessing Research Proposals and Projects of Russian Biology Institutes and the Committee on Dual-Use Technologies, Export Control, and Materials Protection, Control, and Accounting are involved in U.S. nonproliferation efforts. Many think tanks and nonprofit organizations also provide grant money, outreach and education programs, and expertise in support of U.S. government efforts to prevent proliferation of BW capabilities from former Soviet states.

Implementation Issues

The U.S. effort to prevent proliferation from the former Soviet biological weapons complex is similar, in many respects, to its effort to prevent proliferation from the nuclear weapons complex. In both cases, the weapons complex suffers from degraded security systems that remain from the Soviet era,[58] the degradation of the physical plant at the facilities, a lack of income for weapons scientists, reported struggles between hard-liners and reformers in former Soviet states, and the presumed desire of terrorists and hostile nation-states to obtain the weapons capabilities remaining in both complexes. However, there are significant differences between the biological weapons and nuclear weapons complexes and these differences give rise to significant differences in U.S. nonproliferation programs.

Absence of U.S. Biological Weapons Complex

One key difference between the nuclear and biological nonproliferation programs is the lack of a U.S. equivalent to the former Soviet biological weapons complex. The U.S. BW program closed during the Nixon administration. Whereas scientists in the U.S. nuclear weapons complex were logical counterparts for former Soviet nuclear scientists, it has taken time to identify and recruit suitable American counterparts for BW scientists. Furthermore, whereas a sophisticated understanding of nuclear weapons technology was readily available to guide U.S. efforts to prevent proliferation of nuclear weapons from former Soviet states, U.S. understanding of biological weapons technology was more limited. Through BW nonproliferation efforts in former Soviet states the U.S. has been learning about Soviet/Russian BW technology and many observers believe that this knowledge has enhanced U.S. national security.

[58] Kitfield, James. "Nuclear Nightmares." *The National Journal*. December 14, 2001.

Mixed Jurisdiction

When addressing concerns with Russia's nuclear weapons facilities and weapons complex, the United States could work almost exclusively with Russia's Ministry of Defense or Ministry of Atomic Affairs (MINATOM). In contrast, many agencies in the Soviet government shared jurisdiction over the Soviet biological weapons complex. In Russia today the Ministries of Agriculture, Health, Industry, Science and Technology, Defense, and other agencies claim jurisdiction over various aspects of the BW complex. In some cases, the United States can develop projects with individual BRPCs and individual scientists at the grass-roots level, without interacting with any Russian government agency. However, mixed jurisdiction can complicate efforts to dismantle facilities or destroy weapons technology. The United States must negotiate an implementing agreement with the government agencies that own and control specific facilities and technology before it can act directly to dismantle or destroy property. This process is difficult to complete when more than one agency claims jurisdiction.[59] Therefore, biosafety, biosecurity, and dismantlement projects at BRPCs are funded indirectly, through the ISTC. Some argue that this is an inefficient mechanism for managing these kind of projects.

Dual-Use Technology

Most of the technology required for development and production of nuclear weapons is unique to the weapons complex. But the technology associated with biological weapons is also used in medical research, pharmaceutical production, and public health work. Therefore, according to some, if the United States helps fund non-military research using biological technologies, it might risk unintentionally subsidizing a BW program.

Several factors might mitigate the risk posed by the "dual-use" nature of biological weapons technology. First, before beginning cooperative projects with the United States, BRPCs must provide assurance "that they will abstain from offensive research or proliferation activities."[60] Second, ISTC/STCU personnel audit most projects. Third, U.S. agencies require U.S. participants to make on-site visits to BRPCs to ensure that the projects are meeting their stated objectives. In addition, the National Security Council has mandated that all U.S. projects involving former Soviet/Russian BRPCs must be reviewed by the Nonproliferation Interagency Roundtable.

Coordination of Interagency Efforts

According to government officials, the Proliferation Strategy Policy Coordinating Committee, or PCC, which is chaired by a National Security Council senior director, provides the interagency oversight and policy implementation of all U.S. nonproliferation assistance to the states of the former Soviet Union. This committee has representatives from State,

---

[59] Mixed jurisdiction is not an issue in Kazakhstan or Uzbekhistan, where the United States has negotiated implementing agreements directly with the governments of those states.

[60] GAO/NSIAD-00-138

Defense, Energy and other concerned agencies. This committee has several working groups, or subcommittees, that are designed to ensure day-to-day coordination among programs in different agencies. Within this structure, the Nonproliferation Interagency Roundtable (NPIR) reports to the Proliferation Strategy sub-PCC on Bio/Chem Proliferation.

For the past four years, the NPIR has sought to coordinate all U.S. government efforts to prevent BW proliferation from former Soviet states. The Roundtable is chaired by a representative of the State Department and comprised of representatives of various agencies; it reports directly to the National Security Council. The Roundtable helps the various agencies prevent duplication of efforts and seeks to ensure that, together, they address the appropriate priorities. At monthly meetings, members discuss and coordinate implementation issues, such as travel to the FSU or guests arriving from the FSU. When reviewing proposals for collaborative research projects, the Roundtable can reject or mandate revisions in proposals for projects that might reinforce or extend BW capabilities.[61]

The Nonproliferation Interagency Roundtable is adjusting its procedures to keep up with the increasing number of projects. It plans to develop a comprehensive database to facilitate systematic coordination and monitoring of projects.[62]

Measuring the Results

In April 2000 the General Accounting Office (GAO) reported that collaborative research projects involving former BW scientists had "helped to discourage scientists from cooperating with countries of proliferation concern and terrorist groups, while promoting openness at more than 30 former Soviet biological weapons institutes."[63] GAO also reported that about 1,655 former employees of the Soviet/Russian BW program received funding through the ISTC in 1999. Since that time, U.S. biological nonproliferation efforts have expanded as confidence in their effectiveness has grown. Furthermore, these cooperative projects provide the United States with continuous working-level access to Russia's BW sites. Some U.S. participants state that this cooperation allows for better verification than 2-day inspection visits because cooperation involves long-term, routine interaction.[64]

Some observers contend that the United States benefits from these efforts in other ways besides nonproliferation. The research can also address non-weapons related concerns. For example, using a strain of brucellosis from their collection of pathogens, former Soviet weapons scientists are prepared to help the United States produce a more effective vaccine for bison in Yellowstone park.[65]

As of March 2002, more than 40 BRPCs throughout the former Soviet Union have engaged in cooperative projects with the United States. These projects have helped the United States assess the extent to which these facilities were involved in the development and production of biological weapons capabilities. According to State Department officials, approximately 30 of these were deeply involved in the Soviet/Russian biological weapons

61 Correspondence with a State Department official.

62 Interview with a State Department official.

63 GAO/NSIAD-00-138

64 For descriptions of ways in which Russian hosts have tried to conceal BW capabilities from U.S. inspectors see Mangold and Goldberg, *Plague Wars*. pp. 62-140. See also, Alibek and Handelman, *Biohazard*, pp. 137-224

65 "Former Soviet Union weapons scientists may help solve brucellosis," *Associated Press*. October 8, 2001.

program, and therefore, are considered to pose a significant risk for contributing to proliferation. The others were less integrated into the Soviet/Russian BW complex and, therefore, pose less of a proliferation risk.

Some known BRPCs do not participate in U.S. programs; others, such as the State Research Center for Virology and Biotechnology ("Vector") in Novosibirsk, Russia, and the State Research Center for Applied Microbiology (SRCAM) in Obolensk, participate in multiple projects. The Russian Ministry of Defense has denied the United States access to four BRPCs under its jurisdiction.[66] Furthermore, as cooperation has expanded, the United States has discovered additional BRPCs that once participated in the Soviet/Russian BW programs.

Some view the interpersonal relationships built between U.S. participants in these efforts and former Soviet weapons scientists as a significant part of these efforts. In addition to cooperating on projects, U.S. participants and their counterparts in former Soviet states share meals, get to know one another's families, and spend leisure time together. Furthermore, some argue that contact with U.S. companies, NGOs, and government agencies has decreased the social and institutional isolation of Russia's biological weapons scientists. Several people involved in work with former BW institutes and scientists have expressed the view that these interpersonal and institutional relationships are profoundly affecting the decisions weapons scientists are making about how to use their expertise.[67]

ISSUES FOR CONGRESS

The Security Assistance Act of 2001

The Security Assistance Act of 2001 (S.1803) has several sections that are relevant to U.S. policy regarding BW capabilities in former Soviet states.[68] The following discussion is based on Senate action, only. Section 304, entitled "International Nonproliferation Export Control Training," gives the President authority to provide education and training to foreign personnel to enhance their ability to implement export controls that might contribute to nonproliferation activities. Section 305 extends the Soviet Scientists Immigration Act of 1992 and increases the number of scientists that can be relocated under that act from 750 to 950. Some people advocate inviting former weapons scientists to come to the United States to work, as a way of preventing proliferation of weapons expertise. They argue that the U.S. economy and U.S. biodefense efforts would benefit from the expertise of these scientists. Some who oppose this approach argue that it would cost less money to employ a scientist in a former Soviet state than in the United State Some also argue that it is in the U.S. national interest to invest in the science and technology sector of former Soviet states and support their transitions to market economies.

---

[66] They are the Institute of Virology at Sergeyev Posad (formerly Zagorsk), the Center for Military Technical Problems of Biodefense in Ekaterinaburg (formerly Sverdlovsk), the Kirov Institute of Microbiology in Kirov, and the Kirov-200 Institute in Strizhi.

[67] Interview with a U.S. contractor and with several officials of U.S. government agencies.

[68] The Senate passed S1803 by a voice vote on December 20, 2001, and the bill was referred to the House Committee on International Relations as of January 23, 2002.

The Russian Federation Debt Reduction for Nonproliferation Act

Subtitle B of the Security Assistance Act of 2001, the Russian Federation Debt Reduction for Nonproliferation Act of 2001, is designed to encourage Russia to increase its own budget allocations for nonproliferation programs. Russia and other former Soviet states have supported BW nonproliferation efforts in the past by providing infrastructure, staff, and funding to organizations such as the ISTC and STCU. They also contribute to cooperative projects by providing funding or in-kind resources, such as equipment and utilities, to the BRPCs. On the other hand, many experts believe that Russia and the other former Soviet states have sometimes pursued activities that are inconsistent with U.S. nonproliferation goals. For example, a recent CIA report charged that, "During the first half of 2001, Russian entities remained a significant source of dual-use biotechnology, chemicals, production technology, and equipment for Iran."[69]

The Russian Federation Debt Reduction for Nonproliferation Act grants the President the authority to reduce Russia's Soviet era debt to the United States, through a variety of mechanisms, if Russia uses the funds that would have been applied to debt service for projects designed to reduce the risk of proliferation from Russia's nuclear, chemical, or biological weapons complexes. In a further effort to shape Russia's policies on nonproliferation, the legislation states that the President cannot implement the "debt for nonproliferation swap" until Russia makes "material progress in stemming the flow of sensitive goods, technologies, material, and know-how related to the design, development, and production of weapons of mass destruction and the means to deliver them to countries that have been determined...to have repeatedly provided support for acts of international terrorism."

Supporters of this legislation argue that Russia's debt creates such a burden to the state that Russia would be unlikely or unable to devote its own funds to nonproliferation projects while servicing its debt. Furthermore, they argue that this legislation might provide Russia with an incentive to place a higher priority on nonproliferation because it could expect relief from some of its debt. Others, however, note that the United States does not hold much of Russia's debt, so any effort to link nonproliferation with debt relief would have to win the support of other nations, such as Germany, who hold a larger portion of the debt.

The Non-Proliferation Assistance Coordination Act of 2001

The Non-Proliferation Assistance Coordination Act of 2001 was introduced in late 2001 and incorporated into the Security Assistance Act of 2001 as Subtitle C. It seeks to address congressional concerns about interagency coordination of all U.S. nonproliferation efforts in the former Soviet States. It establishes an interagency committee that will monitor U.S. nonproliferation efforts in the former Soviet Union and coordinate U.S. policy with respect to the implementation of those efforts.

When considering this legislation the International Security, Proliferation, and Federal Services Subcommittee of the Senate Government Affairs Committee heard testimony on November 14, 2001, that emphasized the potential benefits of improved coordination of

---

[69] U.S. Central Intelligence Agency. *Unclassified Report to Congress on the Acquisition of Technology Relating to Weapons of Mass Destruction and Advanced Conventional Munitions: 1 January Through 30 June 2001*, January 2002. The Russian Foreign Ministry disputed the charge stating that "Russia strictly meets its international obligations to control the export of sensitive trade and technology." Bellaby, Mara D. "Russia Rips CIA Report on Technology." *Associated Press*. February 7, 2002.

nuclear nonproliferation programs, in general. In contrast, some experts stated that interagency efforts to prevent proliferation of biological weapons were "well-coordinated."[70] Although few experts have focused on the implementation of the biological weapons programs, most who do tend to agree with this assessment. The General Accounting Office reviewed the programs in 2000, and the Bush administration reviewed them in 2001 and neither identified coordination among them as a problem.

Funding

Congress and the Bush administration have demonstrated continued and growing support for U.S. efforts to prevent proliferation of biological weapons from former Soviet states. Upon completing a detailed review of U.S. nonproliferation and threat reduction assistance to Russia and the other former Soviet states, the Bush administration identified the "Redirection of Biotechnical Scientists Program" as one of four programs to be expanded.[71] In the National Defense Authorization Act for Fiscal Year 2002, Congress approved the President's request for $17 million for DOD's efforts in biological weapons proliferation prevention in Russia.[72] In the emergency supplemental appropriations bill passed after the September 11 attacks, Congress added another $30 million "for the purpose of supporting expansion of the Biological Weapons Redirect and International Science and Technology Centers programs, to prevent former Soviet biological weapons experts from emigrating to proliferant states and to reconfigure former Soviet biological weapons production facilities for peaceful uses."[73] Furthermore, in its 2000 report on the BW nonproliferation programs, published before the anthrax attacks of October 2001, the GAO estimated that the United States would spend around $220 million on BW nonproliferation between 2000 and 2004. Most experts agree that this amount may increase now that attention has focused more sharply on the threat of biological weapons proliferation from Russia.

Some experts argue that the United States should increase its efforts to stem proliferation from the former Soviet Union. In January 2001 a panel headed by Lloyd Cutler and former Senator Howard Baker found that, "Current nonproliferation programs in the Department of Energy, the Department of Defense and related agencies have achieved impressive results thus far, but their limited mandate and funding fall short of what is required to address adequately the threat."[74] Some also point out that opportunities to redirect former BW scientists, to secure dangerous pathogens, and to dismantle BW equipment continue to arise as U.S. access to the former Soviet/Russian BW complex expands. Furthermore, according to several U.S. government officials, recent increases in funding for existing programs, while welcome, has been surpassed by an increasing number of viable project proposals.

---

[70] Leonard Spector emphasized the need for greater interagency coordination of nonproliferation programs in general. However, speaking of bio weapons he said, "Through a *well-coordinated* interagency program to create non-defense employment opportunities for former Soviet BW scientists, the Clinton administration successfully engaged a number of former Soviet BW sites in Russia, Kazakhstan, and Uzbekistan," Senate Committee on Government Affairs. November 14, 2001.

[71] White House Fact Sheet, December 27, 2001.

[72] P.L.107-107.

[73] P.L.107-117

[74] Risen, James. "Nuclear Items Sold by Russia To Iran Pose An Obstacle." *New York Times*. January 11, 2001.

ALTERNATIVES FOR THE FUTURE

U.S. assistance to Russia's BW program has sought to support the transformation of the Soviet/Russian BW complex into a set of consolidated, safe, secure, and essentially transparent institutions. Many experts continue to support this objective, but now believe that it will take a long-term investment rather than a short-term crisis intervention.[75] The number of BRPCs known to have been part of the Soviet/Russian BW infrastructure continues to increase. Furthermore, U.S. experts are debating the best long-term strategies for preventing BW proliferation from former Soviet states. Many American participants in U.S. BW nonproliferation efforts consider that long-term nonproliferation goals might be achieved without sustaining U.S.-funded biosecurity and collaborative research projects for the remaining lifetimes of former BW scientists. But some also contend that premature disengagement from BRPCs that continue to pose a proliferation risk is not acceptable.

In recent months, U.S. government agencies and NGOs involved in these efforts have intensified their focus on moving BRPCs towards consolidation and self-sustainability, which many believe is the best way to maximize the nonproliferation benefits that can be achieved by U.S. efforts. Some overall approaches include drawing a new generation of scientists into the BRPCs and preparing BRPCs for external investment. Experts argue that a separate strategy for disengagement will have to be developed for each BRPC, because each one poses unique problems which require unique solutions.

Some analysts argue that the United States should not help the BRPCs become self-sustaining organizations. They argue that market forces will eliminate the BRPCs in former Soviet states because they cannot compete with other biotechnical research and production centers throughout the world. Others, however, counter that without international intervention, market forces might unite former weapons scientists with proliferating states and sub-state groups. Some observers consider the biological weapons complex in former Soviet states to be too isolated and underfunded to safely dismantle itself without U.S. investment. Such observers contend that the goal is not so much to make the BRPCs competitive with other BRPCs throughout the world as it is for the United States to successfully compete with proliferating groups and nation states for the BW capabilities that exist at these BRPCs.

[75] "It became clear that any meaningful result required a more extended and creative kind of engagement..." Civilian Research and Development Foundation. *CRDF 1998-2000 Program Report*, p. 2.

Table 1. U.S. Programs to Assist Russia with Biological Weapons Nonproliferation

Agency	Program	Program Objective
Department of Defense	Cooperative Threat Reduction Program, Biological Weapons Proliferation Prevention (BWPP)	Redirect BW scientists through collaborative research; improve safety at Russian BW facilities in preparation for collaborative research projects; improve security at Russian BW facilities by consolidating and restricting access to pathogens; eliminate BW infrastructure and equipment.
Department of Energy	Initiatives for Proliferation Prevention	Redirect BW scientists through collaborative research; incorporate industry partners to identify market-driven projects that might produce commercial products and results
Department of State	International Science Centers Program (ISTC/STCU)	Provides grant funding to redirect BW scientists to non-military research; provides support for the development, management, and auditing of projects sponsored by other U.S. agencies
Department of Agriculture	Agricultural Research Service (ARS)-Former Soviet Union Scientific Cooperation Program	Redirect BW scientists through collaborative research on diseases that might affect plants and animals
Department of Health and Human Services	Biotechnology Engagement Program (BTEP)	Redirect BW scientists through collaborative research on public health problems
Environmental Protection Agency		Redirect BW scientists through collaborative research on environmental damage caused by biological weapons
Department of Commerce	Special American Business Internship Training Program (SABIT) Business Information Service for the Newly Independent States (BISNIS)	Facilitate business training and exchanges

INDEX

D

G

H

M

N

O

P

Q

R

S

T

U